Why Has
AMERICA
STOPPED
INVENTING?

DARIN GIBBY

NEW YORK

WHY HAS AMERICA STOPPED INVENTING

BY DARIN GIBBY
© 2012 Darin Gibby. All rights reserved.

ISBN 978-1-61448-048-8 Paprback
ISBN 978-1-61448-049-5 Ebook
Library of Congress Control Number: 2011929416

Published by:
MORGAN JAMES PUBLISHING
The Entrepreneurial Publisher
5 Penn Plaza, 23rd Floor
New York City, New York 10001
(212) 655-5470 Office
(516) 908-4496 Fax
www.MorganJamesPublishing.com

Interior Design by:
Bonnie Bushman
bbushman@bresnan.net

In an effort to support local communities, raise awareness and funds, Morgan James Publishing donates one percent of all book sales for the life of each book to Habitat for Humanity.
Get involved today, visit
www.HelpHabitatForHumanity.org.

For America's Inventors

TABLE OF CONTENTS

PREFACE

O ne day while sitting in my office, sorting through an unintelligible set of rejections crafted by a patent examiner bent on making sure that my client would never receive his patent, I decided that I'd had enough of being a patent attorney. "Nothing gets patented these days," I grumbled. It was true. The allowance rate for software patent applications had plummeted to an all-time low of 12%. Even worse, I couldn't remember the last time I'd seen a groundbreaking invention come across my desk. Needing to vent, I wandered down the hall, asking my colleagues to cough up the latest really good idea to venture into our law firm. Try as they might, nobody seemed to be able to recollect any recent ingenious inventions.

Frustrated, I went home and started digging. It didn't take long for me to discover that today Americans invent less than half of what they did 150 years ago. That was shocking.

I wanted to know why. And so I started on a quest. As I began reading about the great inventors of the nineteenth century and their momentous patent battles, the reason for America's lackluster performance with innovation immediately became apparent. And it had a lot to do with the demise of our patent system. Out of that experience came *Why Has America Stopped Inventing?*

Along my journey, I was fortunate enough to meet Jillian Manus who soon became my literary agent. With the help of Penny Nelson, our little team came up with a way to tell this story—one that I hope will convince both America and Congress that life in America could be vastly different … if we really began to invent.

Introduction

THE PATENT GAME

O ver the past two decades, I've met with hundreds of inventors, exuberant as they present to me their new discoveries. Before capitalizing on their ideas, they've been told they need to secure their patent rights. So they've come to my office, toting a prototype or, most often, a few drawings hastily sketched on a napkin. They're nervous and excited and eager to get started. But all that is about to end. The truth is that if they don't have the stomach to put $30,000 on a roulette number in Las Vegas, they'd better not try to take on the patent system. Their odds are better in Vegas. Taking on the modern U.S. patent system to secure a patent is an incredibly long shot. And, if they want to launch their product into the marketplace, they're bound to discover the patent system can present them unseen risks as well. My clients stand a good chance of being hauled into federal court for infringing somebody else's patent— even if the patent office grants them their own patent.

When these entrepreneurs venture into my office, they know nothing about the path that lies ahead. Perhaps they've already attempted to get funding, maybe spoken to a bank or venture capitalist, and gotten advice that they need to file a patent application before any funds will flow. Before handing over cash, astute investors want their investment protected, and they're right when they say that the only way to protect an invention is with a patent. But that's where all the trouble begins.

1

A few years ago, two potential clients came to see me with a bicycle that had an LCD screen mounted on the handlebars. A half dozen sensors positioned on the bike collected data and transmitted it wirelessly to a minicomputer hidden beneath the LCD screen. Sensors on the pedals told if the rider wasn't pulling up hard enough on the backstroke, or if one leg was doing more work than the other. A seat sensor measured the rider's weight, and an incline sensor measured the grade. With a complicated algorithm, the device could compute how much energy the rider expended, down to the last calorie.

Being a rider myself, I immediately recognized the value of this idea. In real time, a rider could tell whether his form was off and how much energy he was expending, and using a GPS device could even produce a map. With the green wave, the weight loss wave, and with cycling coming into vogue, it was certainly a marketable product. I told them how much I liked the idea, but that was the end of the good news. I was about to explain the patent process.

To get a U.S. patent, a registered patent attorney is a necessity. The patenting process is far too complex for a novice. Once the patent attorney understands the idea, he prepares a patent application and then files it with the patent office. That takes about a month. The application then sits for two to three years until an examiner actually looks at it. When he does, he's going to reject it—because that's his job.

How does this work? Typically, the patent examiner will come up with some argument that he thinks the idea is obvious because bikes have been around for over a century, similar sensors can be found on treadmills, GPS systems are in widespread use, and other mapping software already exists. There is nothing inventive about combining known technologies. This forces the patent attorney to argue with the examiner about why he's wrong—why this is really different from any other cycling system. The examiner has rules he's supposed to follow when giving out a blanket rejection, but he usually doesn't.

So there's a long, drawn-out battle between the examiner and the patent attorney, which often involves traveling to Washington several times. Eventually, the examiner decides that you've paid your dues and he'll give you a patent with a few narrow claims. Not enough to keep your competitors out of the market, but something to appease you.

How much and how long for all this? Typically between thirty and forty thousand dollars and three to seven years. During this waiting period, the inventor can mark the product with "patent pending," but that can't stop somebody from knocking off the bicycle's computer and sensor. "Patent pending" just means your application is sitting in the patent office. You can't sue on a patent application. You need to wait until it becomes a patent—on average, five years later.

After the patent issues, an infringer can be sued, but there's more the inventor should know about a patent infringement lawsuit. If the cost to obtain a patent is astronomical, the cost to enforce it is in another galaxy. Plan on at least two million, and that's if everything goes smoothly. Unfortunately, that's how much it costs to sue an infringer. And that's on the cheap side. It can easily go to seven million. Some go as high as twenty.

Most inventors assume that they can get a law firm to take the case on contingent fee, but those cases are rare. If a law firm is going to spend seven million of its own money, it'll want to be compensated for taking that risk. The partners will want at least a threefold return on their $7 million investment, around $20 million. If the attorneys get a third of the recovery, the infringer needs to have had profits of at least $60 million. That's profits, not sales. Few inventions generate those kinds of numbers, and therefore, few patent infringements tempt a law firm to gamble on a contingent fee. For the most part, the legal system works for only the big players. The little guy always gets left holding the bag.

So with all the hassle, time, and money involved, do inventors really need a patent? Absolutely. They'll never get funding without it, and if they don't have a patent, I can almost guarantee they'll be knocked off. If they have a really good exit strategy, maybe one of the big boys will buy them out and foot the bill for the patent infringement suit. But one thing is for sure: If they don't have a patent, they're going to get dinged on the offering price.

And that isn't the end of the patent game. If these entrepreneurs actually roll out their bicycle computer system, they could be sued for infringing somebody else's patent. Most people don't understand that even if you have your own idea patented, you can still get sued for stepping on somebody else's patent.

By now, my clients are beyond frustration and I feel terrible. Repeatedly delivering this kind of news is not what I expected to do with my life. They *do* have a great idea. It's genuinely useful, encourages exercise, and has "green" written all over it. But they're up against the U.S. patent system.

At this point, my clients always ask the same question: "So what can I do? I'm here because I need your help."

For years the only thing I could tell them was to go lobby a member of Congress. That advice came across as less than useless. And it frustrated me to give it. Why couldn't I—a well-seasoned patent attorney—help the very people who made our economy the greatest in the world's history?

Then, a few years ago, I began to wonder whether it was always this way. Was it always this hard to invent in America?

What I discovered was a fascinating saga involving the most famous players in America's history. From 1790 to 1915 their involvement with patents consumed a significant part of their lives. And these weren't closet stories—they were the most popular stories of their day—publicized in the greatest papers and journals. In a world without baseball, the Internet, and Hollywood, millions faithfully followed these legal battles. Patents played such an important role in Lincoln's legal life that he picked two of the nation's best patent attorneys to act as his Secretary of State and Secretary of War. And to Jefferson, patents were so important that he spent his first two years as Secretary of State examining every patent application, even urging Eli Whitney to hold the course with his cotton gin.

The other thing I discovered is that patenting has always been a messy business. America has always struggled with how to protect its inventors.

And they deserve to be protected.

I had little appreciation for what these early inventors went through, the years of persistence and enormous personal sacrifices. Inventing wasn't just a hobby—it was a true obsession, an overwhelming desire to turn an idea into a working reality. Inventing was their religion—their ultimate passion. In my mind's eye, I can see Goodyear languishing in debtor's prison, visited by his wife, who totes a block of rubber so that her husband can keep his dream alive.

But just like my own clients, America's first inventors soon learned the painful reality that finishing an invention was just the beginning of yet another long and arduous journey. Their efforts immediately shifted from inventing to stopping copiers. And because America's foray into the world of patents was truly an experiment, it was inevitable that her first inventors would be nothing short of guinea pigs. Whitney tried sixty patent infringement cases before he finally tasted victory. But by then, his patent had already expired.

Whitney wasn't alone. Other early inventors faced the same roadblocks. And they pleaded for help to stop the infringements. Eventually it did come, in the form of men like Daniel Webster who argued the Great India Rubber case for Charles Goodyear while still Secretary of State, and Edwin Stanton defending Manny against McCormick's mechanical reaper patent after Abraham Lincoln was relieved as local counsel. Although dismissed by Stanton at the eve of trial, Lincoln was so impressed by Stanton's arguments in defending Manny against McCormick's patent that Lincoln later asked Stanton to serve as his Secretary of War.

And it was at this same time, when confidence in the patent system began to take hold, that America experienced an explosion of innovation, a time like none other in the world's history. The period from the 1830s to the 1850s was America's golden age of invention. A time when the legal, political and social climate was prime for promoting innovation. And the inventors that emerged from obscurity were as well regarded as the original founding fathers.

Since then, America has found itself in a slow state of innovation decay. Layer upon layer has been added to our patent jurisprudence, making it next to impossible to obtain or enforce a patent. The result? Ordinary Americans don't invent anymore.

Yet the past holds hope. However flawed, we once did have a patent system that worked. Just look at Singer, Colt, and McCormick. Even at the trailing edge of the golden age of invention when the patent system began to come under attack, it still worked for the Wright Brothers and permitted Henry Ford to avoid infringement of the infamous Selden automobile patent. We can learn from the past and do better. Much better.

Today, we have inventors that are willing to follow the path blazed by Whitney, Goodyear, and Ford. I know them. But we can't leave them out in the cold. We need to reinvent how to patent in America.

Chapter One

LIFE COULD BE BETTER

Pick a thoroughfare through any major city in the world—Fifth Avenue in New York, the Champs Elysees in Paris, Knightsbridge in London, Market Street in San Francisco, Avenida Atlântica in Rio de Janeiro—and if the weather's nice, you'll see thousands of people strolling along, listing to their portable music player, chatting on cell phones, tapping out text messages, or reading email. Occasionally, a runner may zip past, checking her distance and speed with a GPS device strapped to her wrist. Most are oblivious to airplanes whizzing overhead or the street traffic—sleek modern vehicles carrying passengers who surf the web on their PDAs, watch videos on their mobile phones, or slug out a home run on a Nintendo. Some cars even talk to their drivers, directing them to their programmed destinations. It looks as if we're in the golden age of technology, a time like none other in the history of the world. Some say there's even too much, that we're in a state of technology overload.

But today's impressive technological gadgets blind people to the fact that more, much more, is yet to be invented—world-changing technologies that we've yet to see. Because of the accumulation of technologies over the last several centuries, we've become desensitized, fooled into thinking that we are the greatest inventive generation of all time. But we have lulled ourselves into thinking we are greater than we really are. Somewhere yet to be invented are groundbreaking technologies that would make today's gadgets seem trivial. We're so complacent with what we have that no one realizes what our lives *could* be like … if Americans really began to invent.

Imagine the first part of the twentieth century—a time without television, microwave ovens, cell phones, computers, satellites, or the space shuttle. That world seems so different from our own that we tend to think of it as belonging to a type of Dark Age. Yet most of today's technological marvels are small improvements over what was already invented by then. By 1900, our nation's inventors had already produced the steam engine, the train, the automobile, the telegraph, steel making, photography, the typewriter, dynamite, the telephone, the electric motor, the light bulb, the facsimile, and the phonograph. Why don't we see those breakthroughs today? Inventions of this kind were so significant that in 1843 Henry J. Ellsworth, Commissioner of the U.S. Patent Office, stated that, "The advancement of the arts, from year to year, taxes our credulity and seems to presage the arrival of that period when human improvement must end." Although ridiculed and often misquoted for his intentional embellishment, to some extent his statement has proved to be prophetic.

Nearly three decades ago, a hypnotist came to my high school for an after-school assembly. He placed a dozen students under hypnosis and asked them several questions. The most provocative was futuristic: What would the automobile fuel of the future be? The answer, without exception, was water. While most of us weren't smart enough to understand the chemistry behind this, that the useful fuel was really the hydrogen in the water that needed to be split from the oxygen atoms, somehow in our subconscious minds, running a car on water was a real possibility. While scientifically this is a viable alternative fuel, I'm still patiently waiting for my local gas station to install distilled water pumps. And a water-fueled car isn't the only invention we can't get our arms around. What about real-time language translation, cures for cancer—anything other than radiation and chemotherapy—the ability to store light, a replacement for paper, a bed to cure back aches, a replacement for the light bulb, a way to heat houses without a forced air furnace, toilets without water, and what I want most of all, a car like the one on the Jetsons? At this point, though, I'd be happy for a car that gets a hundred miles per gallon.

The "American Dream" was built on the notion that with hard work and ingenuity, and with the support of a strong economy based on the rule of law, anyone could make it in the United States. Is the American dream still a reality, or is it slipping away? In America, we love to envision ourselves as a forward-looking culture, the culture that gave the world the automobile, the airplane, space flight, the computer, and too many medical breakthroughs to count. Few of us realize that the number of patents being issued on our shores has plunged in recent decades. In fact, today Americans on a per capita basis are granted fewer than half the number of patents issued a hundred and fifty years ago. It is a shocking fact. We've been using the same mode of transportation for over a century. And why are we still using coal, gasoline, and natural gas as our major energy sources? We've evolved from the steam engine, the automobile, and the airplane to what? Those who should be today's inventors prefer to focus on downloadable applications for a smart phone rather than on a car that can run on water. It's no wonder we call our economy a "services" economy. Manufacturing is out of vogue.

Don't get me wrong. I love my smart phone and all my clever apps—ones to track my stocks, give me updates on my favorite teams, tell me the weather, and make sure I never get lost while driving. But these "entertainment" inventions mask the real problems facing America. How can we stop the spending of $1 trillion annually on foreign oil? Why can't we invent a solution to our country's energy problems—issues so significant they could lead to our downfall? It's not a trifling matter.

All of this raises a critical question: Why has America stopped inventing? Why are we unable to continue the tradition of groundbreaking inventions of the kind that made our country great? In recent decades technological development has waned, along with our economic strength. After a century and a half of decline, we should be asking ourselves: how can we make this technology explosion happen again?

The answer may well lie in the lives of the great innovators of the nineteenth century, such as Eli Whitney, Samuel Colt, Charles Goodyear, Isaac Singer, Cyrus McCormick, Samuel Morse, and Wilbur Wright, to name of few. Learning what drove these individuals to invent and then feverishly ward off masses of copiers is a key to understanding how America can revive its innovative spirit.

Few realize that our own computer revolution during the 1980s and 1990s closely parallels the series of events that unfolded during the 1830s to the 1850s. To illustrate the point, consider just five industries that emerged from 1830 to 1860: rubber, revolvers, the telegraph, sewing machines, and reapers. Their counterparts can be found in today's semiconductors, the Internet, smart bombs, telecommunications, and ethanol production. During the 1830s, speculation surrounding rubber as the new miracle material rivaled the speculation during our own dot-com era. Throughout this period, our nation laid thousands of miles of telegraph cables that opened instant communication over long distances. America also had its share of defense contractors, like Samuel Colt, hoping to cash in on foreign wars. Add to that an agricultural revolution where machines such as the reaper changed how Americans produced their food, and the sewing machine radically changed the efficiency of its factories.

But back then, the amount of innovation was on a much broader scale, raising the important question as to why we have failed to replicate that scale of innovation. One significant difference is that the vast majority of inventors during the nineteenth century were ordinary individuals working alone, often on farms or in shops. That all changed a century later when researchers flocked to the safety of corporations. It was so much easier to take a regular salary, enjoy a fully stocked research lab, and have the corporation fight the patent battles.

The good news is that with such close technology parallels, it is possible to return to America's golden era of innovation. But to appreciate how this can happen, we first need to understand what happened to Eli Whitney.

Chapter Two

How America's Innovation Began— Eli Whitney

A s the eighteen century waned, the South was in deep trouble. They needed a new cash crop—something they could export to Britain. Without this, many plantation owners faced certain ruin. With no available work, there was talk among even Southerners about emancipating the slaves. Indigo was no longer in demand, tobacco had raped their soil, and an oversupply had sent prices plummeting. Large tracts of land went uncultivated.

The situation isn't far removed from what America faces today. Though we may not spend our money to import manufactured goods from Britain as the early Americans did, we do send our treasure to countries that don't like us so that we can fuel our insatiable appetite for energy. We too are in deep trouble.

But back in America's early days, there was hope for the future—and it had to do with cotton. The craze for cotton started in the 1780s when wool and flax went out of fashion in England. The new material in vogue was cotton, made available by English mill owners who developed a way to mechanically spin and weave the cotton into cloth.

The frenzy also reached America when Alexander Hamilton in 1791 released his *Report on Manufacturers* and speculated that the cotton textile industry could be brought to America, especially after a man named Samuel Slater escaped England with intimate knowledge of how to build the weaving machines used by the cotton textile mills in England. Of course, all of America believed they were innovative enough to replicate these spinning machines.

Regardless, whether in England, America, or both, cotton *was* going to be the next big thing for America's planters.

While Southerners sensed the looming demand for cotton and smelled the sweet scent of money that came with it, they faced one major problem. The cotton England wanted—a kind that was easily cleaned—didn't grow in the American south.

In 1790, there were generally two varieties of cotton: long-staple and short-staple. The long-staple, Sea Island variety, was easy to clean, but it grew only in limited locales, particularly in America's northeastern coastal regions. Short-staple cotton, also called green cotton, could be grown anywhere—it grew like a weed. But the problem with the upland, green cotton, was that it had sticky seeds that were almost impossible to remove. It took a slave an entire day to clean a single pound. The economics didn't work. For this reason, the total amount of cotton produced in the U.S. during 1791 was a mere two million pounds.

But Eli Whitney was about to change all of that. His cotton gin could efficiently remove the seeds, exponentially increasing America's annual cotton production to nearly a hundred million pounds in just twenty years. Never in the history of the world has one change in technology impacted the world's economy in such a rapid manner.

How Whitney ended up in a position to invent the cotton gin is a tale in itself. Whitney grew up as a tinkerer. As a teenager, he lugged a pail of his own hand-made nails from his family farm near Westboro, Massachusetts in hopes of secretly peddling his wares without his father's knowledge. Yet as he reached adulthood, Whitney sensed a bigger future than sticking around the family farm. At age nineteen, Whitney decided he needed to move on and approached his father about attending college. At first his father was reluctant, perhaps because of the cost associated with a higher education. But eventually his father relented and agreed to let Whitney attend Yale College.

Following graduation, Whitney faced the dilemma of most college graduates: He still didn't know what he wanted to do with his life, but he did need a job. His student loans were coming due. Whitney initially secured a teaching position in New York, and when that fell through, he took the only thing he could find: a job as a private tutor for a bunch of rich kids—"gentlemen's" kids—who lived on an estate in South Carolina. Whitney took the job reluctantly because he viewed the South as merely a place with an unhealthy climate. But at least it would help him repay his father and give him some time to read as he considered his future and the possibility of a career in law.

The plan was for Whitney to sail to New York, then take another ship from New York to Savannah, where he would meet his new employer. Arrangements had been made to travel with one Phineas Miller, Catherine Greene, and her five children. This must have excited Whitney as Catherine Greene was somewhat of a celebrity. She was the widow of Revolutionary War General Nathaniel Greene and was returning to her Savannah plantation—a reward for her husband's service during Washington's Valley Forge campaign. She herself had also served beside Washington on those bitter nights, attending to the needs of the soldiers. As Whitney would soon learn, Mulberry Grove, like most Southern plantations, was floundering. The job of keeping the plantation afloat fell to Miller who was hired by Nathaniel Greene to be his children's tutor. After Nathaniel Greene's death, Ms. Greene asked Miller to stay on to oversee management of the plantation.

When Whitney arrived in New York, he encountered a man covered with small pox. Fearful that he would break out during the voyage, Whitney consulted with Miller and Ms. Greene, who suggested that Whitney get inoculated before leaving. This Whitney did, with Ms. Greene patiently looking after Whitney for two weeks while he recovered. When Whitney was finally ready to travel to Georgia, he began to stew over his future lost wages due to his illness. Ms. Greene sensed Whitney's anxiety and offered to pay his fare. That would be the first of many payments she would provide to Whitney on her way to becoming his benefactress.

Upon their arrival in Georgia, Whitney was in for another surprise. He was informed that his tutoring position was no longer available. Not to worry: Ms. Greene offered to let Whitney stay with her until he got back on his feet. The arrangement was somewhat awkward and embarrassing for Whitney. He—a small town Puritan—was staying with an older widowed woman on a large Southern plantation. He didn't pay rent and wasn't really expected to do anything. Ms. Greene, however, made Whitney feel important, and he felt obligated to help out as he could, inventing a knitting frame for her and toys for the children.

It was at Mulberry Grove that Whitney learned of the cottonseed problem. It was the talk of the South; solving it could save them all from a slow and agonizing death. When several neighbor planters were at her home discussing the problem of the seeds, Ms. Greene offered up Whitney's assistance. He can "do anything" she said, probably thinking about the creative toys he'd created for her children.

Whitney could hardly turn down the challenge from his benefactress, though he had never even seen a cotton boll. Phineas Miller sent him down to a basement room where Whitney went to work. What is remarkable is that ten days later Whitney emerged with his first prototype—one that would do the work of fifty pickers. To add to the greatness of the feat, Mulberry Grove was isolated and Whitney had few tools or materials at his disposal. Most of his time was spent building the tools he would need to make the cotton engine.

The basic idea behind the cotton gin was to create a rotating cylinder with a series of teeth. An iron guard with narrow slits was placed over the cylinder so that the teeth projected through the slits. The cotton was fed over the rotating cylinder so that the teeth would tear the seeds from cotton as they became trapped against the guard. The seeds then fell into a box positioned below the cylinder. To remove the cleaned cotton from the teeth, Whitney used a second cylinder covered with a brush to sweep the clean cotton from the teeth on the first cylinder.

Crude as it was, he'd done it—invented a machine that could revolutionize the nation's economy. He knew it. Catherine Greene knew it. And Phineas Miller knew it.

What to do next? Miller, Ms. Greene, and Whitney struggled with the situation. Miller, the businessman, saw dollar signs. Ms. Greene, the socialite, wanted to show it off. Whitney was just plain scared and confused. All the talk made him nervous. Like nearly every other inventor, the moment Whitney realized what he'd invented, he became paranoid. He was certain that somebody was going to steal his idea. Whitney's reaction wasn't unique. Most inventors feel the same way—that they've invented the most revolutionary invention that world has ever seen, and that all of humanity is already conspiring against them. Whitney's initial reaction was to lock the gin up so that nobody could see it. Later, his business plan would reflect this same sensitivity. Instead of commercially selling gins, or even licensing them, Whitney's initial plan was to build regional processing plants where the planters could bring their seeded cotton for processing.

None of Whitney's initial ideas to keep his gin secret worked. It was like discovering gold and trying to hide it beneath the floorboards—a scenario in which only a Silas Marner could find satisfaction. And it wasn't realistic. Almost immediately, the secret was out. Some of it may have come from Ms. Greene's friends that she invited to Mulberry Grove to see the contraption. Whitney, guest as he was, could hardly deny them entrance. Later, there would be break-ins, and soon everyone knew how Whitney's gin operated.

Beyond feeling paranoid, Whitney felt other surges of emotion, those common to nearly every inventor. The creation of something new brings with it not only a sense of pride, but also a notion that this idea is going to change the world. And if it's that important, the financial reward isn't far behind. Unrealistically or not, most inventors feel their invention is going to make them millions—no, billions. Never mind whether the idea is another heated windshield wiper or self-cleaning toilet seat. The emotions are the same. It never hurts to dream big. Whitney was no exception. He wrote to his father that, "Tis generally said by those who know anything about it, that I shall make a Fortune by it." And the more the three discussed their futures and the role that Whitney's cotton gin would play, Whitney was "now so sure of success that ten thousand dollars, if I saw the money counted out to me, would not tempt me to give up my right and relinquish the object."

But unlike most inventors, Whitney was dead on. People *would* try to steal his invention. And it really *was* worth millions, if not billions, in today's dollars.

In the days following Whitney's invention at Mulberry Grove, there was talk of the new Federal law for patents, one that Whitney needed to become intimately familiar with. Then came the need for a business plan, and the need for financing. Whitney didn't know where to turn. Miller took the first step, proposing a plan to commercialize the idea. Miller would put up the money and cover all expenses in return for a half interest in any profits. For his part, Whitney was to immediately leave for Philadelphia to meet with the Secretary of State to procure his patent, and then get to making gins.

This, of course, meant that Whitney was expected to meet with Thomas Jefferson, one of the most famous men in America. From his rural Puritan upbringing, Whitney seemed ill prepared for a meeting with such an esteemed statesman. But his seven months at Mulberry Grove had changed him. Ms. Greene, well accustomed to the presence of men of prominence, took him under her tutelage, teaching him Southern dignity and manners.

Whitney left for Philadelphia on May 27, 1793. He sailed from Savannah to New York, then took a stage to Philadelphia. From Whitney's expense book, we know that he boarded the ship on June 10 and paid $36.25. By June 14, he was on the stage where he spent another $4. But it doesn't appear that Whitney ever met with Jefferson in person, possibly due to Jefferson's schedule. That wouldn't happen for another seven years. Instead, on June 18, Whitney purchased a pamphlet entitled, *Laws of Congress*, for thirty-three cents. On February 21, 1793, Congress had passed its second patent act, making drastic changes to the first Patent Act of 1790. From the pamphlet, Whitney learned what it would take to receive a patent. Whitney discovered that although patent applications were no longer substantively examined and that essentially every patent application was automatically granted, several formal requirements still had to be complied with before the patent would issue. The lack of substantive examination by Jefferson and his Patent Board would end up plaguing Whitney for the life of his patent—fourteen more long years.

After reviewing the requirements, it was clear that Whitney would be unable to meet them while he remained in Philadelphia. Whitney would need to return to New Haven to prepare a detailed write-up of his cotton gin, along with a corresponding set of drawings. And, most important, Whitney would need to submit a working model of his new gin.

Still, on June 20, 1793, Whitney formally lodged his request, paying the requisite $30 fee and his petition for letters patent. He personally addressed his letter to Jefferson, "humbly" requesting "an exclusive Property" in his machine for ginning cotton. His application claimed that the gin could, with one or two persons, clean "as much cotton in one day, as a hundred person could cleane in the same time with the gins now in common use." Then he requested that "your Honor to Grant him the sd. Whitney a Patent for the sd Invention or Improvement, and that your Honour cause Letters Patent to be made out."

Confident that he could soon complete the remaining requirements, Whitney left for New York to face his other problem: How to commercially produce hundreds of gins and complete his patent model. He spent several weeks purchasing the necessary materials and on July 8 paid $4.50 for a train ride to New Haven.

It took Whitney until mid-October before he'd completed his patent specification and drawings. In a letter dated October 15, 1793, Whitney again wrote to Jefferson, including his description, drawings, and sworn oath of inventorship. Whitney would have preferred to pay Jefferson a personal visit so that he could demonstrate his gin, but a yellow fever epidemic had just broken out in Philadelphia, making that impossible. Though Jefferson stayed put in the city during the outbreak, Whitney thought better of venturing into such contagious environs. In a letter to Jefferson, he explained the situation surrounding his efforts to complete the application:

> It was my intention to have lodged in the Office of State a description of my machine for ginning Cotton, immediately after presenting my petition for an exclusive property in the same; but ill health unfortunately prevented me from completing the description until about the time of the breaking out of the malignant fever in Philadelphia.

Whitney explained that he would have sent the materials sooner, but he was worried that business was so disrupted in Philadelphia that his package wouldn't reach Jefferson. Although the yellow fever was still prevalent, Whitney didn't think he could delay any longer, so he resorted to sending the materials by post.

> It has been my endeavor to give a precise idea of every part of the machine, and if I have failed in elegance, I hope I have not been deficient in point of accuracy. If I should be entitled to an exclusive privilege, may I ask the favour of you, Sir, to inform me when I may come forward with my model and receive my patent.

Whitney's letter prompted a quick reply from Jefferson—for he immediately recognized the significance of Whitney's invention. Not surprisingly, the letter, dated November 16, 1793, came from Jefferson's personal residence in Germantown. In it, Jefferson agreed that Whitney had met all the formal patent requirements, except for Whitney's model. Once that was received, Whitney would be granted his patent. Then came the real reason for his letter:

> As the state of Virginia, of which I am, carries on household manufactures of cotton to a great extent, as I also do myself, and one of our great embarrassments is the cleaning of the cotton of the seed, I feel a considerable interest in the success of your invention for family use. Permit me therefore to ask information from you on these points, has the machine been thoroughly tried in the ginning of cotton, or is it as yet but a machine in theory? What quantity of cotton has it cleaned on an average of several days, & worked by hand, & by how many hands? What will be the cost of one of them made to be worked by hand? Favorable answers to these questions would induce me to engage one of them to be forwarded to Richmond for me.

His postscript adds: "Is this the machine advertised the last year by Pearce at the Paterson manufactory?"

On November 24, 1793 Whitney replied, making sure to address every point raised in Jefferson's letter. Whitney explain how he had invented the gin in ten days while staying in Georgia, but that it had taken longer to make a commercial model because of lack of materials. He apologized for not having a gin to sell him, but he assured Jefferson that it was fully possible to make one small enough for family use, and that prices had not yet been decided. He told of his plan to build a gin on Greene's plantation, one big enough to be drawn by a horse. Finally, he told Jefferson that the Pearce gin was not one of his, and that as far as he could tell it was one of the old designs that didn't work, at least not on his new principle. Whitney concluded by telling Jefferson that "It is my intention to come to Philadelphia within a few weeks and bring the model myself" so that Jefferson could personally see how it worked.

Despite his best efforts, it would take Whitney until February 1794 to finish the model. When Whitney finally did arrive in Philadelphia, he was too late. Jefferson was done with Hamilton, had resigned his post as Secretary of State, and was now relaxing in Monticello. So Edmond Randolph, one of the original patent board members and the current Secretary of State, granted Whitney his patent.

Whitney was elated. As he wrote his father,

> I have just returned from Philadelphia. My business there was to lodge a Model of my machine and receive a Patent for it. … I had the satisfaction to hear it declared by a number of the first men in America that my machine is the most perfect & the most valuable invention that has ever appeared in this Country. I have received by Patent.

Whether Whitney understood it or not, by reverting to a registration system Jefferson had greatly reduced the value of Whitney's patent. Instead of having a patent application that was thoroughly examined by a trained patent examiner, Whitney's application was essentially rubber-stamped, leaving its validity in the hands of the court system. Whitney would be one of the first to discover how ill-equipped America's court system was in determining the validity and scope of a patent's claims. And this battle through the court system would nearly ruin him. As he was about to learn, Southern courts felt no need to uphold a federal patent that had never been examined. The new statute left the question of validity to them, and they had no intention of ceding that much power to a Northerner, not when this invention was so vital to their own economy.

Today we too sense something on the horizon … a way out of our economic crisis, a way to stop spending our treasure on foreign oil. Yet the question remains: Where is America's Eli Whitney of the twenty-first century? And perhaps the more important question is this: If he comes, will America help him succeed?

Chapter Three
THE U.S. GETS HER FIRST PATENT OFFICE

America's first patent office ... wasn't really a patent office. It was more like Thomas Jefferson's drawing room.

Whenever a knock came, Jefferson, prone to migraines, must have grumbled and rubbed his temples at the thought of yet another unannounced visitor to his lodgings in Philadelphia, hoping that it wasn't a runner with more patent files.

This thriving merchant city offered little to Jefferson, a man of the earth. He was in Philadelphia, the temporary home of the new federal government, mostly out of a sense of duty. In 1792 Jefferson was well into his term as Secretary of State, fighting with Alexander Hamilton over national fiscal policy and carefully following the disputes between England and France.

Jefferson must have reluctantly taken the files, a stack of patent applications and a few patent models. Moonlighting. It was Jefferson's second job—as if being Secretary of State and fighting with the likes of Hamilton wasn't enough. He added them to the pile, not sure when they again would get his attention.

A year before, it was another story. His enthusiasm for promoting innovation was one of his top priorities. Examining patent applications was not intended to be a mere respite from his foreign affairs. No, if anything, granting patents was even more worthy of his attention than debating over whether the United States should have a national bank. And Jefferson wasn't alone. For the Founding Fathers, protecting inventions was so important that the first patent act placed full responsibility for granting patents squarely on three men, all members of Washington's cabinet. The country's first Patent Board consisted of the Secretary of State, Thomas Jefferson, the Secretary of War, General Henry Knox, and the Attorney General, Edmund Randolph. Granting so much power to one individual by virtue of a patent grant was thought to have been worthy of nothing less.

That is, until reality struck.

It took less than a year before the members of the "Commissioners for the Promotion of the Useful Arts"—known as the Patent Board—realized they had a problem. Cabinet members don't have time to moonlight.

Initially, the Patent Board met on the last Saturday of each month in the State Department office on Market and 7th. Together, they would read each application and stew over the merits of the invention. Even with all their zeal and enthusiasm, they immediately became bogged down trying to sort out when an invention was "new and useful," as the statute required. It wasn't as if they could simply pull a patent treatise off the shelf to guide them. This was new territory. Eventually, they came to the conclusion that they would not grant a patent if the idea was *a mere change in materials or in form.* Obviously, this vague standard was not much clearer.

Their broodings led only to indecisions, and the piles of unexamined applications began to accumulate. The drudgery of giving up a Saturday evening reading through patent specifications became too much, even for Jefferson. The meetings began to slip. Undaunted, the Patent Board concocted a new plan. Instead of wasting one weekend a month, they would review the applications whenever they could fit them into their busy schedules. In July 1791 Jefferson wrote to Knox confirming their agreement to have runners bring the patent files to their homes. The plan was for each member to review an application, make some notes, then forward the file to the other two members of the Board. But every month, the stacks seemed to get higher, the issues more involved, and their time more limited.

The new "leisure" plan didn't last long before Jefferson was again ready to rid himself of the patent business. And the reason was more than just overtaxed schedules. Jefferson had learned what every patent attorney and every patent examiner learns the hard way.

Patents are boring.

It's one thing to sit around the fire, enjoying some tobacco while discussing wild new inventions. It's quite another to slug through pages of dry text. *"The said axel is welded to the said frame."* It's enough to put anyone to sleep. And as soon as Jefferson grasped that, he wanted out—no more patent applications. Though he never admitted it, examining patent applications was now nothing more than a chore, pure drudgery.

But wasn't it Jefferson who thought patents were so important that they couldn't be relegated to anyone else? Just two years before, the entire Congress was exuberant over America's new patent system. Welcome to the world of patents. Nobody wants to deal with them. Few people can understand them, yet we all somehow know that they play a significant role in the nation's destiny.

Whatever Jefferson was thinking (or wasn't thinking) his frustration with patents would lead to a decision that would prove to be a disaster—one that would not only put Eli Whitney in despair but would send innovation reeling for nearly half a century.

———

Although the overwhelming support for the U.S. patent system is well documented, its doctrinal underpinnings are not. The Constitutional

Convention was nearly finished before anyone ever mentioned patents. It wasn't until August 18, 1787 that James Madison submitted a request to grant Congress the power to protect discoveries. Madison didn't propose how that would happen, just that "discoveries" should be protected. The same day Charles Pinckney of South Carolina put some more meat on the bones. He said that the government should grant patents that gave exclusive rights for a limited time. The final language emerging from a committee on September 5, 1787, included language from both Madison and Pinckney and was adopted into the Constitution without comment. It may be the only language in the Constitution that was passed without any debate. Madison's view in one of the *Federalist Papers* seems to have captured everyone's sentiment: "The utility of this power will scarcely be questioned." Madison further argued that patents were so well founded that they were essentially a right at common law. Nobody disagreed. That patent rights ended up in the Constitution almost as an afterthought proves the point.

It also could have had something to do with the latest toy of the day—the steamship. A boat with no oars and no sails that could plow its way right up a roaring river. No, that was impossible. Or was it?

On August 22, 1787, just four days after Madison suggested that a patent provision be inserted into the Constitution, John Fitch invited some of the delegates to a demonstration. Fitch was going to take these overworked men for their first steamship ride up the Delaware River. When word got out, the entire convention adjourned. It was an event not to be missed.

And when the joyride was finished, nobody doubted that Fitch should be rewarded for such a grand invention. If any member of the Constitutional Committee had doubts about whether to include a patent system in the Constitution, those doubts were now put to rest.

Although U.S. inventors were promised protection by the Constitution, none of that mattered until Congress worked out the details of a patent act. Article I, paragraph 8, section 8 simply gave Congress the right to grant patents.

Initially, Congress didn't act. Perhaps it was because nobody in Congress knew what to put in the first patent act. It was an easy sell to reward an inventor for his discovery. It was quite another matter to hammer out the logistics of how it would all happen. And that debate still remains.

What is an invention?

What does the patent cover?

How long is the patent good for?

Who makes all these decisions, anyway?

The first Congress stalled for three years, until George Washington finally prompted them to action. In his State of the Union message on January 8, 1790 he asked them get to work—Congress needed to encourage "exertions of skill and genius in producing [inventions]." At the end of January, Congress did just that and put the task to a House committee. The emerging bill was first adopted by the House, then amended by the Senate. By March, the House and Senate hammered out a compromised bill, and on April 10, 1790, the final version was approved and signed into law by Washington—before the country even had thirteen states. The bill had only seven sections and the entire statute filled two sheets of paper.

Little is known about how the House came up with its original proposal. There weren't many models to follow. The most likely would be from the British because they also granted patents, but their patent system mimicked their social climate. Getting a British patent in 1790 required more than $50,000 in today's currency. Only Britain's elite ever applied. Outrageous? To early Americans—yes. Today, it happens all the time.

Why so expensive? British officials were so worried about patent quality that the application went through multiple offices, each with its own forms—and its own bribes. It essentially shut down the middle and lower class inventor, who therefore didn't invent. And that's when Britain's brain drain began. Those with a good idea and no money flocked to America to pursue their dreams.

We should learn from the British. Today, we too have a multi-step, multi-year process used to insure "quality." And it costs almost the same. In reality, America has backslid 300 years to where the British found themselves in 1790.

It's likely that the drafters of America's first patent act shunned the British model in favor of the earliest recorded patent statute—the Venetian patent decree of 1474. Those were the glory days of Venice, and the city fathers were eager to protect the prominence of their city and its guilds. Inventive minds from all over Europe were flocking to Northern Italy and discovering "ingenious devices." The Council concluded that if there was a way to protect these inventions, keeping others from stealing away an inventor's honor, that these men would continue to "exercise their genius" and continue inventing, all to the benefit of the commonwealth.

The Venetian patent decree gave the Council authority to grant rights to anyone who came up with a "novel and ingenious" device that was reduced to perfection—enough so that others could use it. The right granted was the ability to prevent others from making devices that imitated and resembled the invention for a period of ten years. If unauthorized copying occurred, the infringer would be required to pay one hundred ducats, and the infringing device would be destroyed.

The beauty of the Venetian decree is its simplicity. The entire document was written on a single page. Its sheer simplicity makes it a highly desirable model. The salient provisions include:

1. A requirement that the inventor build the idea to a perfected state.

2. The idea had to be new.

3. Infringement was determined based on whether others imitated the idea.

4. The term was ten years.

5. Damages were fixed.

6. Injunctions to stop the manufacture or sale of any infringing products issued as a matter of course.

In terms of sheer artistic genius, the Italian Renaissance was arguably the most creative period in the world's history. On the technological side, its match is only found during America's own Industrial Revolution, from about 1830 to 1900.

We don't know for sure whether the drafters of the first United States patent act had a copy of the Venetian decree. If they didn't, it would be surprising. With a few small exceptions, the first U.S. patent laws were nearly identical to those found in the Venetian decree.

The Patent Act of 1790 awarded a patent as long as the idea had not been known or used, and if it was useful. In other words, patents were to be granted based on the standard of novelty, meaning the same thing had not yet been invented. Each application was to be considered by the Secretary of State, the Secretary of War, and the Attorney General to see whether it conformed with the novelty provisions of the statute. The application was to include a description of the idea and include a physical patent model, both to show that the invention had been made and that it actually worked as explained in the description. The model was also required so that others could more easily integrate the technology after the patent expired. Logistically, the bill required the President to sign the patent, then return it to the Secretary of State for his signature and to place the Great Seal of the United States. Once granted, the patent was in force for fourteen years.

If the invention was copied, an infringement action could be brought in district court. As a defense to infringement, the defendant could plead that the patent was fraudulently obtained. If the patent holder won his case, the jury was to "sess" damages and the infringing device would be forfeited.

And how much would all this cost? A basic application started off at fifty cents plus copying costs. All told, the cost to obtain a patent averaged about $4 to $5, plus attorney's fees. For example, John Fitch's steamboat patent cost him $4.39, James Rumsey obtained six steam ship patents for $32.18, and Samuel Milliken paid $16.07 for four patents. Taking inflation into account, the government fees were about one fourth of what an inventor can expect today.

If the underpinnings of the Patent Act of 1790 were so well grounded, then why was it overhauled three years later?

Ask Thomas Jefferson.

Not only was Jefferson disillusioned with examining patent applications, he was now facing a myriad of issues he was ill prepared to handle. Jefferson's frustration with the patent system reached its climax with the steamship problem. Before the adoption of the U.S. Constitution, to obtain a patent an inventor could only petition state legislatures or Congress under the Articles of Confederation. But, after the Patent Act of 1790, the floodgates opened and many inventors who were previously unable to obtain patents now formally sought them. The biggest group of these inventors was developing the steam ship, including John Fitch, James Rumsey, Nathan Read, Isaac Biggs, and Robert Stevens. And, if anyone had the advantage, it was Fitch—the man who had chauffeured the Constitutional Committee up the Delaware. But, without a doubt, there was more than one inventor.

With no clear guidelines on which of them should get the patent, each presented arguments as to why he was entitled to the patent. One argued that the patent should be awarded to the first patent application filed with the Patent Board, another to the first petition filed to a state legislature or the previous Congress, another to the first to invent. Jefferson didn't want to sort through the details, and he wasn't inclined to award the patent solely to Fitch, even if Fitch helped to inspire the inclusion of the patent clause in the Constitution. Instead, the Board punted the issue and *awarded them all a patent*. But giving everyone the patent had the effect of giving no one the patent. Without an exclusive right, investors gave up interest and the technology languished. It wasn't until 1805, after these patents had expired, that Robert Fulton rallied the investors needed to bring the steamboat business back to life. If the patent statute had clearly delineated priority issues so that a single inventor was awarded the patent, commercial steam ships could have been running much sooner, particularly when investors had confidence that their investments would be protected by a patent.

Meanwhile, the number of patent applications only grew...and fast. After the first two years, a total of forty-seven had been granted on ideas like Samuel Hopkins' process for making potash, Joseph Sampson's method for making candles, and Oliver Evan's flour-milling machinery. On the heels of these inventions were Charles Peale's bridge and fireplace, Eli Terry's clock, and Peter Lorrilard's tobacco cutting machine. But there were also 114 more applications sitting in the queue. So, at the end of 1791, Jefferson began working on a new bill, one that would take him out of the loop and speed up the process. Later in life, Jefferson tried to justify his actions, writing that the Board members did not feel that they had adequate time to fairly review each application. But Jefferson's idea would more than just remove him from the picture.

It removed everyone.

Jefferson's new patent bill proposed to entirely eliminate examination of patent applications. Instead, it would revert to a simple registration system. File an application, get it rubber-stamped by some clerk, and you've got your patent.

Congress never considered the bill, but Jefferson still got his way. The provision to eliminate examination was passed in 1793, and it also abolished the Patent Board. Jefferson's days as a patent examiner were over. While Jefferson came up with plenty of his own inventions, on ideas like an improved plow, a hemp brake, and a macaroni maker, he never patented any of them. No U.S. president would patent an idea until Abraham Lincoln in 1849 patented a device to lift boats over shoals.

Under the 1793 act, to get a patent, an inventor simply had to lodge a patent application with the requisite formalities and it was registered. Validity and enforcement were left to the courts. For $30—a significant sum in 1793—anyone could get a patent by depositing a model and the right paperwork with the patent office. From examination by the country's highest government officials to no examination at all, the U.S. patent system took an about-face.

Recognizing the problems that cropped up when multiple people claimed to have invented the steamboat, provisions were also added for applications that "interfered" with each other, meaning they all claimed the same idea. For these, a board would determine who invented the idea first, not who filed first. It was based on the concept of fundamental "fairness," a doctrine that would later handcuff nearly every aspect of the patent system.

But if Americans now had the belief that they would get a patent in a single day that everyone would respect, they were mistaken. If anything, just the opposite happened. Even with no examination, it often took months to get a patent, mostly because applicants couldn't meet all the formalities, including having President Washington sign it.

Why Jefferson proposed to eliminate examination altogether is a mystery. Why not just relegate the duty to another government official? Why not simply create a patent office? Had the importance of patents fallen that far in Jefferson's eyes in just three years? When this venture began, Jefferson was such a proponent of patents that he refused to delegate their examination to anyone but top government officials. Granting this powerful property right was so important to promoting innovation that he wanted to make sure that his hand was in every patent grant. So why the about-face?

Yes, Jefferson claimed that he was so overwhelmed with his duties as Secretary of State that he didn't have time to deal with the mundane patent matters. But was that the real reason? That seems inconsistent with Jefferson's character, simply to give up on an idea because he was bored with it.

Perhaps Jefferson really did know what he was doing when he abolished examination. It's reasonable to conclude that Jefferson understood the limitations of the new government. The country was struggling to survive. He wasn't going to be able to petition Congress for a fully staffed patent office. And even if he could, how was he going to train them? There was no legal precedent to follow, and a patent clerk was not the person to be deciding such important policies. The next best choice was for the judiciary to start vetting the issues.

Jefferson may have also realized that this was like the slavery issue, where the Founders had a gentlemen's agreement to table the discussion for a whole generation. Not that patents were controversial like the issue of slavery, but Jefferson still may have felt that the issue wasn't ripe enough to discuss comprehensively. Perhaps he hoped that granting everybody a patent would, over time, reveal issues that could be addressed in a future patent act. If that's what Jefferson intended, it certainly happened. It would take another forty-six years for Congress to finally figure it out. But figure it out they did. The 1836 act was a godsend for inventors.

But the price of this decision would prove to be enormous. And the person who would pay for most of it was Eli Whitney. Jefferson's riddance of examination would affect Whitney the rest of his life—not so much when he reached Philadelphia in June 1793 to lodge his patent application, but absolutely when he tried to stop the hundreds of infringers. Not even Ms. Greene or Phineas Miller could have prepared him for what lay ahead. In his naivety, Whitney could not have possibly understood how Jefferson's decision would nearly ruin his chances of profiting from his cotton gin.

With no formal examination, Southern courts felt little duty to uphold Whitney's patent. And, with one exception, they never did.

Chapter Four

AMERICA'S LAWS GET BROKEN IN

Whitney had his patent, he had his partnership with Miller, and he was ready to make money.

He wouldn't.

The business plan of Miller & Whitney did nothing but encourage infringements of their patent—and as they would soon learn, no Southern jury was going to champion their cause.

They would hate Whitney. Despise him. Yet they couldn't survive without his invention.

The business plan of Miller & Whitney was to keep the gins to themselves, hid away in regionalized processing plants. Then they encouraged Southern planters to fill their lands with cotton, with the promise that Whitney's gins would take care of their seed problems. Just plant, bring us your harvest, and we'll take care of the rest.

Fine—except that Miller & Whitney wanted forty percent all of the cotton they processed as payment. Looking past the usurious royalty, their plan had some logic to it. At upwards of $500, none of the planters had the cash to buy a gin, so giving up some of their cotton was an easy way to handle the cash problem.

But forty percent was just too high of a rate for the business-savvy Southern planters. Even worse, Miller & Whitney failed to execute. The partnership never came close to being able to process the amount of cotton churned out by the Southern plantations at an ever-escalating rate. Yes, Miller had convinced the Southern landowners to plant, yet he and Whitney now couldn't make good on their promise. Granted, most of it wasn't Whitney's fault. He'd even canceled his trip to England to secure a patent there so that he could personally make more gins. But back in New Haven he caught malaria, then in the spring of 1797, his shop burned down, completely destroying 20 completed gins and all his tools. It would take seven more months, and a generous financial donation from Catherine Greene, before he could build and ship twenty-six more.

It was too little, too late.

Miller's promises left the Southerners in a bind. They had their cotton, but they weren't about to give up two-fifths of their harvest, especially when Whitney couldn't even guarantee that he could process it.

Patent or no patent, they were going to process their cotton. They weren't going to let a monopoly stop them. After all, didn't they just fight a war with Britain over the monopolistic chains she'd wrapped around her thirteen colonies? So the Southern planters made their own gins and challenged Whitney to come after them.

He did.

But both Whitney and Miller far underestimated the difficulties that loomed ahead. The Georgia planters immediately went on the offensive, taking Whitney and Miller by surprise. Not only did they boycott Whitney's gins, but the offended Georgia planters started rumors that cotton processed by Whitney's gins was damaged—inferior goods. The tales quickly spread to England, stopping any overseas demand for Whitney-processed cotton. Now, there was simply no revenue to pay off their ever-increasing loans. Miller had mortgaged Mulberry Grove, Catherine Greene's pride and joy, to the hilt.

With both sides pinned down, a showdown was inevitable. Miller filed the first lawsuit in May 1797 against Edward Lyon, a man who was accused of dressing in disguise as a woman, breaking into the gin room at Mulberry Grove, and stealing one of Whitney's gins. Most likely, Lyon learned of the gin when Miller made the mistake of showing the gin to several local planters. For two years, Lyon ran a brisk business selling infringing gins.

The suit was brought in Georgia while Whitney was busying himself making more gins. Miller supposed Whitney's talents were better used increasing their inventory, rather than fighting legal battles. But Miller was naïve as to how Whitney's patent would be treated. Still, Miller did his best. Miller wrote Whitney about his diligence in making sure they were fully prepared for trial.

> We had the Judge with a Party to dine with us twice before the trial came on and got him fully prepared to enter into the merits of the case. We had also got the tide of popular opinion running in our favor and many decided friends who adhered firmly to our cause and interest. Added to this we got the trial brought on, against every measure they could devise for postponement and found them perfectly unprepared as to a knowledge of the strong grounds of their cause and without a single evidence in their favor. We were on the contrary pretty well prepared and neglected no means to become as much so as possible.

Miller thought he had the case in the bag. But he hadn't considered the strength of Lyon's defense—one that would successfully be borrowed dozens of times. The Southern planters quickly picked up on Jefferson's dismantling of the patent system, and they were fully prepared to take advantage of it.

With no substantive examination, patents carried no presumption of validity, leaving the decision solely in the discretion of the jury. In essence, the jury became the country's patent examiners. Whitney's future was now being turned over to a pack of uneducated, self-interested jurists who hadn't a clue about patents. How could they? America's only experience with examining patents took place in Jefferson's drawing room—and that only lasted a couple of years before Jefferson ended substantive examination of patent applications.

The defenses quickly took shape. The one that would catch hold and be repeated again and again was that Whitney didn't invent the cotton gin; he'd stolen it from others. For every trial, there was a never ending stream of witnesses willing to come forward and testify that they knew of other gins. Even with clear jury instructions from the judge, juries could believe whatever they wanted.

And in Georgia, they did. Whitney brought several patent infringement suits at the end of the 1790's and lost them all.

"The Judge gave a charge most pointedly in our favor," Miller wrote of one case, meaning that the judge gave instructions to the jury favoring a finding of infringement. After the jury instructions, the defendant was certain that he was going to lose. According to Miller, "the Defendant himself told an acquaintance he would give two thousand dollars to be free from the verdict—and yet the jury gave it against us after a consultation of about an hour."

It wasn't a big intellectual leap for Miller to grasp the problems with the 1793 act. "… the imperfections of the Patent Law frustrated all our views and disappointed expectations which had become very sanguine."

What happened to Whitney would set a pattern for how to attack patents for the next fifty years. Simply get the jury to believe that the invention was already made by somebody else. It didn't matter how dubious the evidence. After all, the power was all in the hands of the jury. If the U.S. government didn't want to be bothered with sorting through a patent's validity, why should they?

It was a strategy that worked until 1836 when the patent system was so obviously broken that Congress stepped in and revamped it, forming a patent office that would substantively examine applications. Instead of placing a patent's validity squarely on a jury, patent examiners would once again take the first crack. When Singer was sued for patent infringement and tried to argue the sewing machine was previously invented, his defense flat out failed. Colt's infringers too tried a similar "already-invented" argument and failed, as did Goodyear's infringers.

But at the end of the eighteenth century, an argument that an idea was already invented, whether fabricated or not, still worked. And it worked perfectly.

Miller and Whitney not only didn't expect the verdict, but it essentially threw their financially reeling partnership into bankruptcy. The 1793 Patent Act allowed for triple damages, and Miller had already factored the recovery into his balance sheet. The damages were going to fund further expansion. At least that is what Miller thought. The reality was that he now had to sell Greene's Mulberry estate to pay creditors. The estate sold for a mere $15,000—not nearly enough to cover $38,000 in debts. For Catherine Greene, the silent partner, the woman who had weathered the freezing nights in Valley Forge to bring freedom to her country, the reality of being forced to leave her estate could have been nothing short of utter devastation.

After years of courtship, Ms. Greene had married Miller on June 13, 1796 in Philadelphia. Out of deep respect for Ms. Greene's sacrifice, President and Martha Washington served as witnesses to the wedding. With her marriage to Miller, Ms. Greene was as committed to the partnership as was Whitney himself. Difficult as it was to lose her home, Ms. Greene kept fighting.

Miller thought he could at least argue for a new trial, but he learned that Congress was sloppy when they'd printed the 1793 Act. The wording required an infringer to both make *and* use the invention—as opposed to make *or* use. It was a clear typographical error, yet the word "and" would plague Miller and Whitney for three more years. Lyon could simply argue that although he had made plenty of gins, he didn't use any of them. It was the planters who used them. If Whitney went after a planter, he'd face the same argument in reverse—the planter was using the gin, but he certainly did not make the gin.

This left Miller to conclude that "the people of this state openly declare that they will defy the power of the Federal Laws ... and make no secret of their means of safety in such defiance."

Who could blame them? After all, this was their livelihood, and it wasn't their fault that Congress had left them an opening they could use to enhance it.

When word got out that courts were unwilling to enforce Whitney's patent, it only served to embolden the infringers. Whitney wrote to Robert Fulton that there were three gins within fifty yards of the courthouse, close enough to hear the rattling machinery, yet he could do nothing to stop them.

After constant petitions, Miller and Whitney in 1800 finally convinced Congress to fix their typo. But if he thought that would solve his problems, he was wrong. The infringers had no intention of stopping.

So why did Whitney keep going? Why keep filing fifty-nine more suits? Wasn't life too short? For any rational business person—yes. For inventors—no. It's a trait common to nearly every inventor. It's not a matter of cutting losses and moving on, or in embracing the Christian notion of turning the other cheek and finding solace in forgiveness. Inventors don't think that way. They can't. The invention is a part of them, a creation that's been wrongfully taken from them.

The cotton gin patent was Whitney's baby. And he wasn't letting go, no matter what the price.

His opponents had justified the infringement on the notion that patents were nothing more than illegal monopolies, against which they had just fought a war. Whitney disagreed.

I am well aware that much obloquy has been attempted to be thrown on the principle, as well as the practice of exercising exclusive rights under the authority of Patents—And to give plausibility to such attempts, the justly obnoxious epithet of monopoly (as applied to the force limitation of a general right) has been constantly misapplied. It is not the use of any general privilege which we would confine to ourselves, but the right of managing as we please our own property, a property which can be rendered useful to us in no other way than by sharing its benefits in some stipulated proportion with other members of the community.

Whitney was livid that politicians—especially Georgia's governor—actually encouraged infringement because they believed Whitney's patent was nothing more than an aristocratic monopoly, referred to as a general privilege, that could be ignored because America did not stand for such atrocities. In Whitney's mind, patents were different. They were not a general privilege, but a very limited right given him by an act of Congress.

This same debate still rages today as some feel that patents really are intended to be a "general privilege," akin to a fundamental right. Those in this camp want every sort of protection Congress can invent to protect patents. The result has served only to bog down the patent system, producing volumes of unduly complex statues put in place to protect inventors like Whitney. But the pendulum has gone too far. Not even Whitney would have wanted this. He didn't want a general privilege—only a way to share his invention with the public and be adequately compensated. Three decades later, Congress would finally understand Whitney's point.

But what would happen in the future provided no solace for Whitney. Undaunted, Whitney and Miller continued tracking the offenders in Georgia, suing them in turn: Henry Chambers, Captain John Hunton, James Hutchinson. Sue as they would, Whitney and Miller lost every case.

While the infringers continued in Georgia, Whitney turned elsewhere in hopes of finding more friendly markets, this time with a little more wisdom and a new plan. Based on advice from an old Yale classmate, Whitney chose to license the patent. First up: the entire state of South Carolina. It was a bold move to license an entire state, rather than dealing individually with each planter. He'd be trading a large number of potential licensees and corresponding royalty streams for the convenience of only one. For those who have ever tried to license a patent and experienced the administrative nightmares of keeping licensees in line, it was a well-reasoned decision. South Carolina could pay Whitney a lump sum, then recoup its investment by levying a tax on every gin sold. Even though South Carolina would likely pay a smaller royalty than Whitney could achieve if he negotiated licenses from each planter, an agreement with the state would remove an administrative nightmare for Whitney.

And it was a real possibility. The men of South Carolina treated patents differently than their neighbors to the south. As a colony, South Carolina had actually passed a patent act. They had a tradition of protecting inventions, and it was Charles Pinckney of that state who proposed the language to be placed into the Constitution.

Whitney prepared himself for the upcoming negotiations. He visited Washington where, for the first time, he met with Thomas Jefferson and James Madison, who gave him letters of recommendation. And his perseverance paid off. In 1802 he convinced the South Carolina legislature to take out one license for every planter in the state. Whitney asked for $100,000. South Carolina gave him $50,000. The deal provided $20,000 up front with staggered future payments. In return, Whitney agreed to provide the planters of South Carolina with models containing his improvements and to return any fees he had already collected in the state, a total of $508. Whitney was somewhat disappointed with the license fee, but he couldn't afford to lose the deal. It was $50,000 he desperately needed to repay past debts.

Deal in hand, Whitney took a 2,800 mile ride on horseback, first to Georgia to visit Catharine Greene, then back to New Haven. After that, he visited North Carolina, which now also sought a license. But before he could finalize the agreement, disaster struck.

South Carolina reneged on its deal to pay $50,000. And the state wanted its money back. Several of the legislators had been swayed by the constant badgering from Georgia planters, accusing their sister state of selling out. The Georgians taunted them, cajoling the South Carolinians with statements about how they were paying good money that didn't need to be paid.

Whitney's agreement with South Carolina concerned Georgia planters. If South Carolina went ahead with the deal, it legitimized Whitney's patent and made Georgian's look like the horse thieves that they really were. The agreement with South Carolina could now be used as evidence in Georgia courts to show an entire state believed Whitney had a valid patent.

The heightened concern in Georgia came about when a man named Hodgen Holmes lost an infringement case to Whitney. Because the Georgian court never recorded the judgment, he weaseled his way out. Still, it raised the possibility that Georgia planters could be held liable for infringement.

The case against Holmes was an interesting one. Holmes had his own patent on the cotton gin, one of ten that Whitney counted as having been registered based on his same design. Because of Jefferson's decision not to examine applications, anybody could apply for a patent, regardless of whether the same invention had already been patented. This forced Whitney to sue Holmes "to have his patent vacated,—after a tedious course of litigation & Delay we obtained a judgment on the ground that the principle was the same & that his patent was surreptitious—his patent was vacated & Declared void." The court also ordered Holmes to pay damages to Whitney. "He came forward & paid up the costs & purchased a License of us to use the Machine for which he had pretended to get a patent & we now hold his not given for the License." But when the court clerk neglected to record the judgment, Holmes took the position that the suit was void from the beginning. Nobody in Georgia, including the courts, disagreed.

Now Holmes was resurrected and shipped to South Carolina with his fabricated story about how he had actually invented the cotton gin before Whitney. The legislators took his testimony at face value and yanked the contract, all on the false pretext that Whitney's patent was invalid. It was the same argument being made in dozens of courtrooms throughout Georgia.

Whitney was furious. To bolster his case, he went to Washington seeking affidavits to prove he was the true inventor. Sick and fatigued from traveling in ill weather, he lay in bed waiting to regain his strength. But he couldn't get his patent battles off his mind. "The Cotton Machine is a thing of immense value & by pushing hard I hope to realize something for it … but so large a proportion of Mankind are such infernal Rascals that I shall never be able to realize but a trifling proportion of its value. You know I always believe in the 'Depravity of Human nature.' … God Almighty is continually pouring down cataracts of testimony upon me to convince me of this fact."

After returning to good health, Whitney secured the testimony of several men, including Nathaniel Pendleton, General Greene's aide-de-camp during the Revolutionary War, to state that they had witnessed the machine at Mulberry Grove. Pendleton was so offended by South Carolina's actions that he immediately wrote to General Pinckney. Even with faithful backers, the decade-long battle was taking its toll. Whitney was becoming disillusioned. And Georgia's meddling with his affairs in South Carolina had left a sour taste in his mouth. "I might as well go to Hell in search of Happiness as apply to a Georgia Court for Justice," he wrote.

In 1803, the South and her belligerence finally beat Miller, the hopeless legal battles had worn him down both physically and emotionally. Miller's untimely death made Catherine Greene a widow for the second time. It was "Mellancholly tidings" for Whitney. He must have blamed himself for the strain that had been placed on Miller. The death of his friend meant that Whitney was alone in his battle against the infringers in Georgia and the hostile South Carolina legislature.

Still, Miller's death didn't slow Whitney. He was even more determined to justify his cause. Under the terms of the South Carolina deal, Whitney still needed to complete the improved models and repay his licensees. Back in New Haven, he slaved away until the models were finished. Then he promptly left for South Carolina—still enraged at his ill treatment.

When Whitney arrived in South Carolina in 1804 the authorities threatened to throw him into jail if he didn't repay the $20,000. Whitney openly defied them. Cooler heads eventually prevailed and the deal was put back in place. Many South Carolinians deemed themselves men of honor and thought it reprehensible to renege on a contract—even if it was a bad deal. Now that Whitney had supplied the promised models, these men felt that Whitney should be paid the remaining $30,000. And he was.

Good news usually comes all at once. That was true for Whitney. The same day, he received news of a deal with Tennessee. Now he had three states licensed, a more than tacit recognition of his patent's validity.

All that remained were the ongoing infringers in Georgia. But now he had a bankroll to go after them. The only problem was the looming expiration date of his patent. By 1807, it would be all over—unless he could convince Congress to extend his patent, which they never did. With only a few years left on his patent, any victories would be in name only. He simply couldn't bring enough suits in the short amount of time to collect any significant damages. And everyone knew that his patent—which had never been upheld by a court—was getting ready to expire. The vultures in Georgia were now circling.

None of that meant anything to Whitney. He had licensing money from the two Carolinas and he was ready to litigate to the end. He met defeat after defeat. No jury was willing to uphold a patent that was never examined, especially not with so many testimonials of previous gins that predated Whitney's. The stories were becoming legendary.

Undaunted, Whitney plodded ahead with even more lawsuits. Emotionally he needed to go through this. He needed vindication. To him, it was a solemn truth that "many of the Citizens of Georgia are amassing fortunes, living voluptuously and rolling in splendor by the surreptitious use [of his gin], while I am, and for more than six years past have been, chained down to this spot, struggling under a heave [sic] load of debt contracted for my very subsistence and expenses while I was solely employed in inventing and perfecting this machine."

Session after session, the Georgia courts put off Whitney. Judges kept granting continuances, stalling until the day that Whitney's patent would expire. Still, with three other states taking licenses, Georgia was quickly losing credibility. Their only remaining hope was to fend off the attacks until the patent was gone.

But then in 1806, a year before his patent expired, Whitney drew Judge Johnson. Whitney's ill fortunes would finally come to an end.

Because Whitney faced so much opposition, he decided to go on the offensive. Even if the patent office couldn't give him the presumption of validity, he could do it himself. So he "introduced the testimony of several witnesses, residing in New Haven, to prove the origin and progress of his invention."

Then he readied himself for the parade of evidence that was sure to follow. By now, he'd heard every lie. But this time, they wouldn't invalidate his patent.

First came a model of "a machine used in Great Britain for cleaning cotton, denominated the 'Teazer' or 'Devil'" that was supposedly "seen in England, about seventeen years ago." The next witness "testified that he had seen a machine in Ireland, upon the same principle, which was used for separating the motes from the cotton, before going to the carding machine." Next, the defense raised Holmes—again.

Whitney came out of his corner, fists at the ready. He had his witnesses ready to show he was the true inventor. But Judge Johnson didn't want to hear it. He didn't need to hear it. The evidence he needed was everywhere. Just one look out the window and he could see the impact of Whitney's gin.

> It is not necessary to resort to such testimony to maintain this point. ... There are circumstances within the knowledge of all mankind, which prove the originality of this invention more satisfactorily to the mind than the direct testimony of a host of witnesses. The cotton plant has furnished clothing to mankind before the age of Herodotus. The green seed is a species much more productive than the black, and by nature adapted to a much greater variety of climate; but by reason of the strong adherence of the fibre to the seed, without the aid of some powerful machine for separating it than any formerly known among us, the cultivation of it could never have been made an object. The machine, of which Mr. Whitney claims the invention, so facilitates the preparation of this species for use, that the cultivation of it has suddenly become an object of infinitely greater importance than that of the other species ever can be. Is it then to be imagined that if this machine had been before discovered, the use of it would ever have been lost, or could have been confined to any tract of country left unexplored by commercial enterprise? But it is unnecessary to remark further on this subject. A number of years have elapsed since Mr. Whitney took out a patent, and no one has produced, or pretended to prove the existence of, a machine of similar construction or use.

As a last ditch effort, the defense argued that their gin operated by a different principle—the equivalent of today's non-infringement argument.

This argument failed to catch Whitney flat-footed. Before trial, Whitney had built a replica of the infringing gin. Now, he was able to show that it operated identically to his own.

The judge agreed. Then, perhaps apologizing for the pain inflicted upon Whitney by entire state of Georgia, the judge concluded:

> With regard to the utility of this discovery, the court would deem it a waste of time to dwell long on this topic. Is there a man who hears us who has not experienced its utility? The whole interior of the Southern states was languishing, and its inhabitants emigrating, for want of some objects to engage their attention, and employ their industry, when the invention of this machine at once opened views to them which set the whole country in active motion. From childhood to age, it has presented us a lucrative employment. Individuals who were depressed with poverty, and sunk in idleness, have suddenly risen to wealth and respectability. Our debts have been paid off, our capitals increased, and our lands have trebled in value. We cannot express the weight of obligation which the country owes to this invention; the extent of it cannot now be seen.

And so Whitney's victory, at the conclusion of a thirteen-year battle, was little more than a verbal vindication from a federal judge. With his patent was ready to expire, that was about all that he could expect. Georgia had dragged its feet for over a decade, virtually eliminating any chance for him to collect on his patent. When all was said and done, Whitney's recovery was heartbreaking. He finally did get his $50,000 from South Carolina, and another $14,000 from North Carolina. Tennessee's payment is lost to history. Whitney received another $9,500 from various individuals. All told, Whitney received around $90,000. If Whitney had collected even a small fraction of the revenues made possible by his gin, he should have received millions.

Looking back, Whitney was convinced that the reason his patent was intentionally ignored was because it was simply too valuable to the nation's economy. In 1811, Whitney wrote to Robert Fulton that people would "have had no difficulty in causing my rights to be respected if it had been less valuable & used only by a small portion of the community. But the use of this Machine being immensely profitable to almost every individual in the Country all were interested in trespassing & each justified & kept the other in countenance."

But that wasn't the real reason everyone shunned his patent. From a practical point of view, it is always wisest not to be too greedy—a lesson Whitney learned too late. Otherwise, everyone will fight you. It's a common mistake still made by inventors today. If you come up with a great idea, make the licensing fee reasonable or be prepared for a fight.

But shouldn't Whitney have been entitled to do whatever he wanted with his patent? Even if that kept the entire South in dire straits?

Yes, if the patent system were designed correctly. If patents aren't upheld, no matter the cost, patents will have no teeth. In that case, everyone suffers because innovation stops as both inventors and investors realize that they will be unable to get a fair return on their investment. Whitney's patent was largely ignored because his invention was so needed that nobody was willing to go without it. And because of the weak patent act of 1793, there was really no way to uphold his patent. When the Southern states made the decision that they would use Whitney's gin without paying for it, there was little that could be done.

The result? The message was loud and clear—patents weren't worth anything. And the truth was that they weren't. They needed some clout. If you want to promote innovation, you need to craft patent rights so that they set the proper balance between promoting innovation and turning over the technology to the public, and then you need enough guts to enforce those rights. To do that, you need an effective patent statute that requires a thorough examination of the new idea. Ultimately, to promote innovation, the patent system needs credibility, and for that, people need to know exactly what patents cover and whether they are valid. For Whitney, the lack of examination permitted prevaricators to fill the court rooms, with dubious tales of prior cotton gins. Today, the battle ground has moved from prior invention to whether an idea is obvious. Either way, the patent system has lost most of its credibility.

So what of Jefferson's decision to stop examination? Good idea?

If he wanted to see its effect on innovation, he didn't have to wait long. He'd created a patent system that granted patents that were easily shot down and that lent itself to endless fraud. Although Americans did continue to invent, they did so on only a small scale. It would take until the 1840s before the patent system would prove so entirely lacking that Congress knew they had to step in.

In fact, the patent office would have to burn down before that would happen.

Chapter Five

IN COME THE MODELS

When Jefferson dismantled the patent system in 1793 and transferred the burden of examining patent applications onto the federal courts, what was the impact?

For every 800 patents issued in the early nineteenth century, 100 found their way to courts. Compared to today, the rate is at least double. It was a burden the courts were ill equipped to handle. With all the other new constitutional issues facing a fledgling nation, patents were the last thing the courts needed to clog their dockets.

So did anything good come out of the first patent act?

Absolutely.

Even with the strain thrust onto the patent system, there was one bright spot—and that was its patent models. The model requirement was unique to U.S. patent law, one that has never been adopted by any other country.

But it was a gem, and it's an absolute shame that even the U.S. abandoned the model requirement in 1880. In its day, the model requirement was not only a source of great pride for Americans, but served to limit the scope of patents and make sure that patents were awarded on quality ideas. It's a provision we desperately need to resurrect.

What is a patent model?

A patent model is a miniature prototype of the actual product, and its submission was required for nearly a century after enactment of the first patent act. In the language of the patent office: "The model, not more than 12 inches square, should be neatly made, the name of the inventor should be printed or engraved up, or affixed to it, in a durable manner."

As the Industrial Revolution waned, the requirement was eventually retracted. These relics of the past have long since been scattered to the winds, with many finding their way into shelves of today's patent attorneys—paper weights and office decorations.

But a few, like the Westinghouse air brake, are still on display in today's patent office. The Westinghouse model sits on a small wooden platform about a foot long and six inches wide. The components consist of a series of steel tubes for transporting pressurized air to a mechanical brake. The parts are sliced in half, giving it the look of a cross-sectional drawing from a modern patent application.

Is it really fair to force an inventor to build a sophisticated model in addition to drafting a patent application? Wouldn't this by itself stifle innovation? In the nineteenth century, the cost to make the model could often be more than that of the application. And for those inventors, when most were common people—farmers, laborers, schoolteachers, carpenters—requiring a patent applicant to submit a model might easily have been cost prohibitive.

Somehow, the inventors still found a way. Perhaps this was because they were confident that this investment wasn't throwing good money after bad.

How did they make the models? After all, very few had the talent or the tools to whip together a miniaturized version of their inventions.

In America, if there's a way to make money, somebody will step to the plate. For patent models, it was no exception. Dozens of boutique model-making shops quickly sprang up in Washington, D.C. Take an armful of drawings, hover over the machinist for a few hours, and out pops an operating prototype—physical proof of invention, not just some whim of an idea.

The insistence of models by the first Congress likely came about because of a precedent set a hundred years before. When America was made up of thirteen loosely associated colonies, some did allow for patent grants. In those cases, an inventor could petition the colonial legislature or assembly to grant him the exclusive right to use or sell his invention. But legislatures—ever leery of abuses of power—were reluctant to grant any exclusionary rights based on a mere idea. And so they rarely awarded patents.

Smart inventors learned that it was easier to obtain patent rights if they constructed a working model to demonstrate that their idea really worked. Patent models also served a deeper purpose. Legislators knew that patents were granted only if the public would benefit from the idea after the patent expired and everyone was free to use the technology. Models provided an easy way for legislators to see what would eventually be turned over to the public.

A good idea is worth repeating.

And so the first Congress copied this idea. The Patent Act of 1790 required the submission of a miniature model to the Secretary of State.

Jefferson believed that patents were part of a social contract that necessitated government protection, but he also was concerned that government grants of patent rights were subject to abuse. Patent models were a way to curb the grant of excessive rights by ensuring that applicants adequately described their invention so that others could benefit from their technological contribution.

Even when, in 1793, Congress followed Jefferson's prompting and eliminated formal examination of patent applications, they still kept the model requirement. Jefferson felt that models were that important. Thus his insistence that Whitney submit his model before he would register his patent.

When William Thornton was appointed in 1802 to handle patent affairs, he published a pamphlet stating that "if the machine be complex, a model will be necessary" and that "many machines are so complicated that no one but a skillful artist can comprehend their construction or mode of operation without inspection of a model." Under Thornton, models were filed in about eighty percent of the cases. Though not explicitly mentioned by Thornton, another factor in requiring models was that a large portion of the population was illiterate and unfamiliar with technological and scientific publications. Models were a hands-on library for the illiterate. After all, the common mechanic and farmer did the bulk of the inventing during the nineteenth century.

Although Jefferson had rid himself of the patent business in 1793, that didn't mean the logistical issues of handling patent applications went away. Yes, Jefferson and the Patent Board were no longer needed to examine applications for novelty, but somebody still had to register the applications. Patent registration was now purely a clerical function, making sure that all the paperwork and models were properly submitted.

The problem was that there were simply no available bodies to do it. Almost by default, the job fell to the clerks in the State Department who tried to fit in patent registrations with their other foreign duties. In June 1800 the federal government moved from Philadelphia to Washington, D.C. In what was still mostly a mosquito-infested swamp, the State Department took up quarters with the Treasury Department. At the time, the State Department had a total of only eight employees.

With more patent applications and models rolling in, it became clear that the U.S. needed a patent office and somebody to run it. The top choice was James Madison's next-door neighbor, William Thornton. The two had more in common than a fence. Together they raised and raced horses. Since James Madison was Secretary of State, the appointment was an easy one, especially because Thornton had just been relieved of his duties as Commissioner of the Federal City now that Washington had been chartered as the new government headquarters. So on June 1, 1802, Thornton was hired by the State Department to take over its patent responsibilities with a salary of $1,400 a year. Not so coincidentally, that was the same amount generated each year by patent filing fees. Thornton was given one clerk and told to find space as best he could in the State Department building.

And so the patent models came pouring in. By 1807 James Madison, who was still Secretary of State, wrote to Congress indicating that the number of patents was doubling every four years and that the patent office needed more help. By now, Thornton was also acting as justice of the peace for Washington, and he needed another clerk. What the patent office really needed was a building of its own. One visitor to Thornton's room in the War and Post Office building reported that his registration books, along with the accompanying models, were haphazardly scattered about on shelves as well as the floor and were covered with dust. He further observed that the room was far too small for its purpose, and badly in need of a book case.

For his part, Thornton kept pestering Madison about the need for new space, and of his ill treatment as head of patents. He complained about how he was acting as a museum curator for the models, but with no place to put them. And, because of the large numbers of patents being registered, he had no time to write out the specification for each one. To this latter point, Thornton did a great disservice to those who were granted patents. Oliver Evans was the classic example. He'd asserted his patent for a process to mill flour against an infringer in Pennsylvania, but the court said his patent was invalid because the patent office failed to write out the description. He was left without any remedy.

Some things were out of Thornton's control, such as mail delays and the time it took to receive signature from the Attorney General, the President, and the Secretary of State. This was one of the primary reasons why it took Eli Whitney a year to receive his patent.

Thornton continued to pester Madison about his overwhelming workload and the need for a new patent office. He also confided in Madison of his financial predicament due to some poor investments, requiring him to hold up on his farm while he continued registering patents, at a rate of 160 in 1808. When that didn't work, Thornton wrote to President Jefferson informing him that in 1809 the number of patents had risen to 219 and that he was now hiring an assistant at his own expense. Congress finally got the message and on April 28, 1810, authorized the purchase of a new building to be shared by the Patent Office and the Post Office.

Their selection of buildings was an interesting one. President Madison chose Blodgett's Hotel, a building with a sordid past. It was originally constructed in 1793 by Samuel Blodgett, who was a land speculator hoping to cash in if the federal government were moved to Washington. Blodgett took the fortune he'd made trading in East India and started a lottery as a way to advertize his real estate business. The winner of this lottery would receive a hotel worth $50,000. The problem was that the hotel was never completely finished before Blodgett bankrupted. Thornton became a bail bondsman to Blodgett and when Blodgett fled the city, Thornton was left with more than $10,000 of his debts.

The government, needing another federal building, eventually paid the winner of the lottery $10,000 for the hotel. The initial report of the building's condition was dismal. Three families were squatting in the hotel, along with an assortment of hogs. Although the walls were standing, all that existed of the floors were the joists. Nearly every window was broken. Still, they had a building. Work began right away to finish the floors, with the patent office having a receiving room for new applications, an office for the superintendant, and a large room in the center of the building for exhibiting the patent models. Thus, Blodgett's Hotel would become America's first patent museum.

By September, 1810, the building was sufficiently complete for the move. Although cabinets for the models were still being constructed, Thornton was set to move in. To memorialize the event, he wrote a letter to the Secretary of State, which ended up being more of a tirade than a thank you note. Thornton took issue with how patents had taken a back seat. Originally they had to be approved by the Attorney General, Secretary of State, and Secretary of War, and now there were no requirements. Although Thornton didn't have power to refuse registration of a patent, he felt it was his patriotic duty to inform the applicant if the idea wasn't novel and deserving of a patent. Thornton's battle with Robert Fulton when refusing to register the patent on Fulton's steam ship was on direct point with this doctrine. Thornton freely spoke his opinion that applications should be examined for novelty—a battle he would never win.

While Thornton mused over the value of the patent models to America, he probably had little idea that these models would be the only thing to save the last government building from British troops four years later. Tensions between the British and the U.S. escalated into war when Britain began restraining U.S. trade with France, with whom Britain was already at war, along with capturing American seaman and forcing them to labor on British ships. The War of 1812 lasted three years, but was felt most keenly in the nation's capitol during the British raids on the Chesapeake Bay beginning in August, 1814. In what would become known as the "Burning of Washington," America would lose the White House, the Capitol, the Navy Yard, and most other federal buildings. When these raids began, most government officials realized the city could not be defended and fled. To his credit, Thornton decided to stay—and to save the patent office documents. Thornton still had his farm in present-day downtown Bethesda, and he arranged to move the patent office papers there. With his clerk, George Lyon, and model maker, Thomas Nicholson, Thornton purchased wood and iron strapping and hired two others to construct boxes to hold all the Patent Office papers and books. Using hired wagons, they hauled the laden boxes to Thornton's farm. The task took two days, Monday and Tuesday of August 22 and 23. Thornton chose to leave the models in the patent office. There were hundreds of them, and their weight and bulk prohibited their move.

With the documents secured, Thornton joined James Monroe, the Secretary of State, to scout out the enemy. The next day, on August 24, British troops moved through Bladensburg, Maryland with little resistance. Thornton took his family to Georgetown to watch the fires blazing as most of Washington's buildings burned to the ground. The next morning, Thornton learned that Blodgett's Hotel was next on the list. He rushed into the city in an apparent attempt to save one of his own patent models, a musical instrument. When he reached Washington he watched as British troops set fire to the War Office. A British officer, Major Walters, along with 150 soldiers, were awaiting the command from Colonel Timothy Jones to burn the patent office. Thornton plead with the major to spare his musical instrument. According to Thornton's account, the plea went further. Thornton explained that this was the patent office and contained hundreds of scientific models that could benefit the entire world, not just America. He told Major Walters that if he burned the building, he would go down in history as a nefarious criminal, akin to the Turks who burned Alexandria and its world famous library.

Walters was bothered enough by the allegation that together they went to Colonel Jones, who was busy burning Joseph Gales' printing shop. After Thornton again plead his case, Jones gave the order to spare the patent office. A few hours later, a hurricane ripped through the city, extinguishing all the remaining fires and prompting the British to flee the city. The storm was so violent that it also blew off a part of the roof of Blodgett's Hotel. A month later, on the night of September 13, 1814, Fort McHenry was bombarded by the British, with Old Glory illuminated by the explosions. Although Americans now remember that Francis Scott Key wrote the words to *The Star Spangled Banner* during the War of 1812, British seamen went home repeating the story of the monstrous storm that forced them from America's capitol.

With the British gone from Washington City, Congress had a serious problem. They had no place to meet, and the next session was scheduled in two weeks. Because Blodgett's Hotel was the only building still standing, Congress took it over, and they stayed for another year until the Old

Brick Capitol building was constructed on the present site of the Supreme Court Building.

Back in Blodgett's Hotel, the number of patent models continued to grow. Over time, new submission guidelines refined the submission requirements. Rules published in 1828 stated that

> in all cases where it is believed that the nature of the machine will be more clearly shown by it than by drawings alone.... These models should be neatly made, and as small as a distinct representation will admitted; they ought to have a permanent label affixed to them.... They will be carefully kept for the advantage of the patentee and the information of the public.

The submission of models not only served to educate the public, but was critical when the patent was enforced in court. Models were useful to explain the invention to judges and juries who helped decide both whether the patent was valid and whether the accused device infringed its claims. Beyond that, the patent office wanted proof that a mousetrap was capable of catching a mouse. They wanted to know that gears actually turned or that earth was actually plowed.

By 1835 the number of accumulating models had become a serious issue. Henry Ellsworth, the new Patent Commissioner, commented that the weight of the models was threatening the integrity of the building. It was the beginning of many changes that would happen the next year.

Although the displayed models were touted by the country, they soon proved to be a source of fraud as unscrupulous individuals would copy them, then file their own applications on the same ideas. Because of a loophole in the 1793 Patent Act, applications couldn't be rejected even if the same idea had already been patented. Applications weren't substantively examined for novelty. It was becoming clear that America's patent system was rapidly deteriorating. It was due for an overhaul.

The year 1836 was a year where Congress allocated money for a new patent office, and a new patent statute was passed. Under the new law, patents once again began to be examined for novelty, and the new Commissioner of Patents saw the models as being "absolutely necessary" to the examiners as a way to determine whether an idea had been previously conceived. In other words, when determining whether an idea was new, the examiner could look to the collection of previously submitted models to determine whether the idea had already been invented. With a new set of patent laws and a striking new building under construction, the future of America's patent system was bright. But on December 15, 1836 a raging fire tragically destroyed the entire library of 7,000 models, along with all the patent files, including many notable documents, such as Robert Fulton's original steam engine drawings.

The only saving grace was that the new patent office, which today houses the National Portrait Gallery, was already under construction. Only a few years later, the patent office was back in business, collecting more patent models. Over the decades, as patent filing grew exponentially, the new patent office once again became cluttered with thousands of models. Display rooms were set up where millions were dazzled by their genius, ingenuity, and craftsmanship. This display, the Old Curiosity Shop, was a source of pride, not only for the inventors, but for all of America. It was so large it took up the entire third floor of the patent office, with its four grand halls packed with glass cases housing the models. Eventually, guide books were printed and tour guides were offered to steer the patrons through the maze of corsets, railroad machinery, hoop skirts, and paper-making machines, mingled with relics such as Benjamin Franklin's printing press and George Washington's camp equipment from the Revolutionary War.

Chapter Six

AMERICA GETS RUBBER FEVER

When Thomas Jefferson and John Adams both passed away on July 4, 1826—America's fiftieth Independence Day—most Americans firmly believed their timed departures were a divine manifestation. Simultaneously leaving this world on America's most celebrated day was nothing short of a sign from God, his seal of approval, a bold pronouncement of America's grand destiny on the world's stage.

Yet who would step up to steer America to its promised destiny? America needed a new generation of leaders.

It wouldn't be Washington, the man revered as a God for crossing the Delaware in a surprise Christmas Eve attack on the British. It wouldn't be any of the revolutionaries. They were now in their graves, and harrowing times were on the horizon. Only a decade later, America would find itself ravaged by financial speculation and panic. Mob rule reigned. South Carolina was making rumblings about seceding.

New leaders would come, but from the most unlikely of sources. Not from the ranks of lawyers, politicians or the privileged class. Rather, those who would transform America into an economic powerhouse and a world leader were its humble inventors.

59

Help came in the form of America's great innovators, men like Goodyear, Morse, Colt, and McCormick, who forged ahead with their desire to invent, even in the most dire of circumstances. But they too would fight through their own Valley Forges. And as the economic crisis deepened, so did their struggles. In the mid 1830s, these names were so obscure, nobody could have guessed that it would be these inventors who would step forward and save America. But they did.

In their own way, each of these inventors somehow sensed his role in changing America. For Samuel Morse it started with art. Morse was set to place his name in the history books with his brush. His contribution would be to help America remember her past. Through his paintings, he would inspire the next generation to revere America's founders and to do their part in making America great.

In 1825, just a year before the deaths of Jefferson and Adams, Morse could sense that his artistic talents would bring him fame. That year, Morse had been commissioned to paint the portrait of the very man who helped win the Revolutionary War: the Marquis de Lafayette, Washington's most trusted general. Lafayette's services were a gift from France, the result of years of delicate negotiations between Benjamin Franklin and the French court. With the infusion of French soldiers and French gold, Washington was able to fend off the British and Hessian troops.

Now Morse would be the one to create the last great painting of Lafayette. The commission came none too soon. The men of the revolution were quickly disappearing. But for Adams and Jefferson, who would die the following year, all the signers of the Declaration were now gone.

So Morse responded to the summons and traveled from New Haven to New York to meet with the venerable Marquis. The price of the commission was substantial: $1,000. On his way, Morse couldn't help but think this was just a stepping stone. For his ultimate prize was to have his name, his artwork, adorning the Neolithic temple that was now being constructed in Washington. It was a dedication to the great men of the Revolution, those who had founded America, whose images would soon fill the walls of the Rotunda. For many, the Rotunda was intended to be America's own Pantheon to the gods, where Washington would one day adorn the ceiling in a great apotheosis, where the revered man would be taken up to live with the gods. If Morse was born too late to be one of the founders, he would find a way to become immortal another way— through his paintings. He'd do it just as DaVinci did in The Last Supper, or Michelangelo in the Sistine Chapel, or Raphael with The School of Athens in the Vatican. Morse may well have believed that his art could preserve the spirit that created America, just as former artists inspired future generations with the spirit of the Renaissance.

But Morse would never realize his dream of a painting in the Rotunda, or even finish his portrait of Lafayette. Tragedy would call him elsewhere, to redirect his path to a more urgent need. Before Morse could finish Lafayette's portrait, he received the tragic news of his wife's impending death. By the time he reached home, she was already buried, leaving him with three children. Life would soon get worse for Morse. Only a few months later, he would lose both his father and mother and fall into a deep depression.

Yet Morse would find fame, and he would end up in the Rotunda. Not as the artist of one of the paintings, but as the subject. In fact, he appears there twice.

And Morse wouldn't be the only one to have his image there. Except for Washington, the figures who would adorn the capital's ceiling wouldn't be America's founders. Rather, it would be a new set of revolutionaries who would take America to her envisioned destiny. Not by laws or political will, but by technology. Instead of achieving independence from European powers by a prolonged war, this new generation would strengthen America by inventing. America would produce more men like Whitney whose cotton gin had given both farmers and merchants a taste of the wealth and self-sufficiency that came with innovation.

Although the Founding Fathers constituted some of the world's greatest legal and political talent who came together to create the legal framework for America, their contributions alone did not guarantee America's future success. The great innovators of the nineteenth century would step in and play an important role in ensuring America's economic development.

And the sacrifices of these men are comparable to those made by the signers of the Declaration. For not only would they invent America out of its crisis, they would end Jefferson's experiment and rewrite the laws vital to protecting this innovation. In a sense, it would be a new Bill of Rights, an expansion of the Constitutional mandate to promote the useful arts and sciences. The ensuing explosion of ideas, inventive genius, and technological economic development would prove to be the only way for America to assert herself as a leader of the free world.

But in 1825, Morse knew none of this. Life had dealt him a severe blow. To pull himself out of despair, Morse took a sabbatical to Europe, concentrating on his painting, studying the artwork in the Louvre, and meeting with James Fennimore Cooper in Italy. The journey had shades of Ben Franklin's ventures a half century earlier, escaping to London, then to Paris for his own intellectual stimulation. Although the visit to Europe revived Morse, he constantly longed for his three children back in America. He was frustrated by the slowness of the mail. Morse couldn't understand why it took months for a single letter to reach his children.

That would soon change. On Morse's return home from Europe in 1832 on the *Sully*, one lively dinner conversation turned to the recent discovery of electromagnetism and how a French physicist had developed a system that used an electromagnet to move a needle to different letters. When Morse's dinner companion, Charles Thomas Jackson, a scientist fascinated with electromagnetism, confirmed that electricity could travel down a wire, Morse had his idea for transmitting signals over long ranges.

Yet Morse still clung to the hope of painting his masterpiece in the Rotunda of the Capitol. After all, he was a painter, not a scientist. That was when John Quincy Adams did him, and the world, one giant favor. In 1834 the last four paintings in the set were put up for commission. Adams, who headed the House committee responsible for the selection, wanted to attract additional artists and extended the opportunity to foreign painters. Protests rolled in, including one from Morse's friend, James Fennimore Cooper. The letter was published in the New York Evening Post and was attributed to Morse. Outraged, Adams yanked Morse from the list of potential candidates, and virtually ended his painting career. Morse decided to put all of his efforts into completing the telegraph.

Lacking funds to support himself, in 1835 Morse took an unpaid position teaching art at New York University, hoping to earn money from tutoring. Even with little income, Morse worked nights and weekends until his first telegraph machine showed promise.

Morse's first receiver consisted of an old picture frame fastened to a table, with the wheels of an old wooden clock being used to move the marking paper forward. A wooden pendulum holding a pencil moved over the paper to make marks each time the circuit was broken. Morse showed this to one of his colleagues, Professor Leonard D. Gayle, who helped Morse with his first attempt at a long distance transmittal over 40 feet of wire. Even with limited funds, Morse managed to improve the distance to 1,700 feet, using wires strewn across campus. By October 1837, that distance was up to ten miles of wire wound around reels set up in Dr. Gale's lecture room. Morse was convinced the design was good enough to file his patent application on the now famous electromagnetic telegraph. His first filing was known as a "caveat," a mere place holder meant to protect his basic idea until he filed a full-blown application. Morse's next challenge would be how to commercialize the telegraph.

━━━━━━━━━━━

And while Morse's passion for painting ultimately led him to the telegraph, it would be Samuel Colt's passion for sailing that would inspire him to invent the revolver. Colt's original dream, beginning while a student at Amherst Academy, was to be a great explorer on the open seas. That, he believed, was what would make his name immortal. So in 1831, he hired himself out to the *Corvo*. While there on the open seas, Colt, like Morse, experienced his flash of genius. Mesmerized by the locking action of the ship's wheel, Colt realized that this same mechanism could be used in a firearm. Colt conjectured that lifting the gun's hammer could rotate the breech until one of its chambers (that held the bullet) came in line with the barrel. While still on this voyage, Colt carved his first prototype out of wood.

When Colt returned to land, he knew he had a good idea. But he had no idea how to make a revolver commercially, let alone market it. There was also the question as to whether unscrupulous opportunists would copy it—just as they did with Whitney's cotton gin. Most important, Colt needed money to build a line of firearms. Even though America had plenty of investors looking to invest, they were reluctant to invest in Colt's idea. Colt didn't have a patent, and the venture capitalists of the day were unwilling to take a risk on Colt.

So how does a person raise money in the 1830s? In Colt's case, by drawing further on his creativity. In one of the history's stranger stories, Colt became an entertainer, taking the stage name of Dr. Coult and embarking on a laughing gas tour of North America, where he educated the public on the virtues of inhaling nitrous oxide. The money he raised he hoped to use to make more firearms.

Colt left on his fundraising tour on March 30, 1832. For the next three years he took his show from the Carolinas all the way to Nova Scotia, where he charged fifty cents admission while typically shelling out a $5 rental fee plus costs for lighting. Overall, Colt raised close to $1,300, which he paid gunsmiths to develop an assortment of prototypes.

The entire venture was rather juvenile, a trait Colt never seemed to outgrow. To capture his listeners' attention, Colt developed a pat script, explaining the virtues of the "exhilarating gas." To some, he explained, it made them ludicrous, encouraging them to leap and run. For others, the effect was to put them in involuntary fits of laughter and fill them with "pleashourable sensations."

But there were dangers as well, Colt warned his audience (or perhaps these were mere theatrics), and they needed to dispose of their knives and weapons before inhaling to guard against an accident. With all the disclaimers out in the open air, Colt himself inhaled and the show began.

Humiliating or not, the show enabled Colt to commission ten pistols, seven rifles, and one shotgun from a number of gunmakers. By August 1835 Colt had enough prototypes to consider his next step—a visit to the patent office.

While Morse struggled with an invention to bring the world together through a series of dots and dashes and Colt experimented with one to strengthen the American army, Cyrus McCormick was stewing over how to harvest wheat using a machine. Yet McCormick would take a different path than both Morse and Colt. McCormick, inheriting his father's passion for machinery, had for decades been contemplating the problem of efficiently harvesting grain.

Before McCormick came on the scene, grain had been harvested the same way for millennia. The stalks were hand cut with sharp scythes. Following the reapers were gatherers, who would gather the cut grain and bundle them into sheathes, where the grain was allowed to dry. The main problem with using scythes was their inefficiency. On average, a man could cut less than three acres of grain a day. With America's exponentially growing population, the number of workers required to produce America's food would become impractical. In 1830, three-fourths of America's work force worked in agriculture. If that trend continued, by 1870 the number of agricultural workers would be nearly 29 million. Much of the required labor was attributable to harvesting crops by hand. And perhaps the most time consuming of all, was the harvesting of grain.

In 1831, McCormick was determined to find a solution to this problem, and he did so by taking up his father's failed twenty-year project to construct a mechanical reaper. The clanking machinery on his father's

532-acre farm near Lexington, Virginia, with its own grist-mill and distillery, routinely landed Whitney's father the derision of his neighbors. But where his father failed, McCormick was determined to succeed. Using every spare minute, McCormick began his tinkering. He *would* make his father's harvester cut grain.

The 1831 prototype cut a mere six acres of oats—sort of. Much of the grain was trampled to the ground. What McCormick had been able to figure out was a way to bring the grains to his reciprocating blades so that they could be cut. Eventually, McCormick came up with the idea of using a reel to gather the stalks, then deliver them to the blades. He also used a divider to isolate the stalks that were to be cut. As the horses pulled the reaper across the grain field, the cut stalks fell backward onto a platform where a man stood and gathered it into sheaves.

After two more years of hard work, McCormick was cutting ten to twelve acres a day. Still, McCormick refused to file his patent application. He was a perfectionist, and the reaper wasn't performing up to his satisfaction.

McCormick's delay let the competition slip in. In the spring of 1834 McCormick learned of an ex-seaman, Obed Hussey, who had been marketing his own reaper for almost a year. Hussey, born in Nantucket, was raised to be a whaler. Like Colt who also loved the sea, Hussey was a tinkerer, loving to sort through mechanical problems. Hussey's venture in the reaper business came about because of a dare. He was working on a machine to mold candles when a friend challenged him to invent a machine to reap grain because that would make him a fortune. Immediately, Hussey gave up on candle making and a year later had his reaper.

Outraged at this slight, McCormick rushed to the patent office and filed his own application on June 21, 1834, then warned Hussey that he had built a working reaper previous to 1833. The two would end up in a decade-long feud. Still, McCormick refused to take his reaper to market for another five years—until he was certain it was worthy of selling. Then, he would guarantee farmers could cut at a rate of 1.5 acres per hour, a five-fold increase over the 3 acres per day now done by hand.

———————

Although in the early 1830s the telegraph, revolver, and reaper were still essentially unknown to the world, rubber was not. Rumors of this new "super material" shot through the financial markets, sending inventors scurrying for a chance to catch the wave. Almost overnight, America had caught rubber fever.

Rubber was like nothing America—or anyone else in the world—had ever experienced. Boots that could keep your feet dry, even in rushing water? A waterproof rain jacket? It was unheard of. Yet at the end of the 1820s, it all seemed within the realm of possibility.

Although latex, the sticky white fluid from which rubber is made, had been known for centuries, the ability to turn it into rubber was a new phenomenon. Americans had experienced "sticky fingers" when picking dandelions, but the significance of latex didn't occur to anyone until the "rubber tree" from northern Brazil came to the world's attention. It was this latex that could most easily be turned into rubber. And Brazil had enough trees to supply any world demand. Make a few incisions in the bark, put a cup at the base of the tree, and wait for the milky white substance to drizzle away.

Latex became popular in Europe in the 1700s when it was spread over the fabric to produce a waterproof cloth. The latex could be turned into a paint if it and pigments were combined with turpentine. But it wasn't until the 1820s that Brazilians began making rubber shoes and importing them to America. New England imported eight tons of shoes in 1826. Four years later that number jumped to 161 tons.

And that's when the craze began. Ordinary Americans starting investing in start-up rubber companies with a reckless abandon. No matter that they knew nothing about the material itself—they unwisely thought they could make immense profits when every foot in America cried out for a rubber boot.

Rubber companies came out of the woodwork, and their bank accounts were bursting with invested money. The numbers were staggering, just like those of the dot-com start-ups that would raise millions overnight. Take the Roxbury India Rubber Company. In 1833, its founder—a leather worker who wanted a way to make leather waterproof—raised $30,000 to start the company. The amount of capital raised by the company ballooned to $240,000 by 1836. In today's dollars, that figure would be in the tens of millions. Most of the rubber companies sprang up in Massachusetts, although Connecticut, New York, and New Jersey had their fair share.

And that provided an opportunity to Charles Goodyear. Goodyear, of slim build, clean shaven, and with coal black hair, found comfort in experimenting with his rubber samples, hoping to find a way to make them immune to temperature changes. If he found any spare time, that was spent with his large family. Like McCormick, Goodyear was also raised on the family farm—on the Naugatuck River in Connecticut. Also like McCormick, Goodyear's father was a tinkerer—inventing scythes, clocks, pearl buttons, lamps, and a spring steel fork for lifting hay. At age twenty-five Goodyear took his wife Clarissa to Philadelphia to open a hardware store. But Goodyear was no business man. After only a few years, his hardware business failed. Goodyear used this as an opportunity to proudly tell Clarissa that he was born to be an inventor. He was going to fix the "rubber problem." And she believed him—she always believed in him.

Goodyear's spark of interest in rubber came in the fateful year 1834 when he invented a new valve for a rubber life preserver, then tried to market the idea, only to discover—along with everybody else—that rubber in its current form was useless. When the summer temperatures elevated in 1834, the shoes and hats melted. Left out in the sun, the goods turned into a gooey, stinking mess, just like caramel or taffy, but with a smell worse than manure. Customers returned their pungent shoes en masse, demanding full refunds. Following the hot summer months of 1834, $20,000 worth of claims were posted with Roxbury India Company alone. Almost overnight, the rubber companies disappeared—and their investors' money evaporated. The capital losses for Roxbury were a whopping $400,000.

The technology failure of rubber made it all the way to the Great Basin. It all happened with a rubber boat made by Horace Day—the man who would haunt Goodyear for decades. When the Salt Lake Valley was nothing but an uninhabited desert, sparsely covered with sagebrush, John C. Freemont watched in horror as Day's rubber boat—hauled all the way from Washington D.C.—began to fall apart in the middle of the Great Salt Lake. Freemont, exploring the West, had also been fooled into thinking that rubber was a miracle material that could be used for anything, including boats. It was nearly a miracle that his men were barely able to row the disintegrating vessel to shore. It was more than a decade longer before Brigham Young would lead his followers to the valley. And it would take that long before Goodyear had the solution.

Still, Goodyear was obsessed with finding a stable, non-melting rubber. With hundreds of thousands of pounds of useless rubber lying around, it was easy for Goodyear to take up his experiments. But with no income, Goodyear was soon thrown into debtor's prison. It was a place he would come to know well. But even in jail, Goodyear continued his quest by convincing Clarissa to bring him a block of rubber and a rolling pin so he could continue with his experiments.

For the Goodyears this was normal life. They bounced from house to house, from city to city, getting evicted and watching their belongings auctioned off by the local sheriff. The places their living would take them were horrific, usually in slums, bordered by whorehouses. Even in those circumstances, the neighbors complained about the stench and noise emanating from the Goodyear residence. A peek inside revealed a rat's nest of rubber samples, scattered pots and pans, all infused with a noxious, vile odor.

If 1834 was bad for investors, it paled in comparison with what would happen in 1837 when a nationwide panic tore through the capital markets. Half the nation's banks failed, and many stocks became worthless. Andrew Jackson, the believer in an agrarian America, was out of the White House, replaced by Van Buren. Political unrest was rampant as those affected by the financial collapsed looked for someone to blame. Many questioned whether this was the end of the Great Experiment. America was on the verge of collapse.

At this same moment, America's innovators were on the brink of making their technological breakthroughs. They needed help. They needed capital, they needed encouragement, and they needed assurances that their ideas would not be stolen. If their ideas were not protected, if they were treated like Whitney, the movement would fizzle. If these men all had one thing in common, it was that they were going to fight to protect their ideas—even if it took every dime of their fortunes.

Fortunately, help was on its way. And it all started with Colt's visit to the patent office.

Chapter Seven

A NEW PATENT OFFICE AND PATENT STATUTE

I n 1835, Henry L. Ellsworth was appointed the new patent office superintendent. He would be a godsend for inventors. For most of his life, Ellsworth knew little about patents. But he did understand complex legal matters. More important, he understood inventors. His father, Oliver Ellsworth, was the Chief Justice. His twin brother, William, was the governor of Connecticut and a classmate of Samuel Morse.

Ellsworth too sought a life of public service, but one with more adventure than either his brother or his father. Before applying for the post of patent office superintendent, Ellsworth's commission took him to the Indian tribes on the American frontier, where his travelling companion was Washington Irving.

What brought Ellsworth to the world of patents undoubtedly had something to do with Ellsworth's close friend and fellow Hartford denizen, Christopher Colt, father of Sam Colt. Ellsworth was soon to become a godfather figure to Colt and likely did more to bring about Colt's success than any other. As fate would have it, Ellsworth applied for the superintendent position in mid-1835, just as Colt was preparing to patent his revolver.

Ellsworth took his post on July 8, 1835, and he began making drastic changes without delay. Although most Americans understood the failure of their patent system, Ellsworth was one who was determined to do something about it. Housecleaning was his first order of business. Organizationally, the patent office was in shambles. Ellsworth spent his first month creating patent files and having them indexed. Then he took the sixty models that were cluttering his office and put them in storage. He also created a comprehensive list of every patent application and made up new patent office letterhead.

It was amidst this chaos, on July 24, 1835, that Colt paid Ellsworth a visit, seeking advice on how to patent his revolver. The past four years had been ugly, utterly humiliating years for Colt. After returning from his laughing gas tour he worked with almost a dozen gunsmiths to complete the design. Then came the rugged life on the road, peddling his guns to any army major who would show an interest. At six feet, Colt was tall, outgoing, and a lively bachelor, and the nomadic life fit him well. Yet he desperately needed another way to raise more money. So Colt decided to put his faith in the patent system. If he could secure his patent, he hoped investors would come flocking in.

The advice Colt received from Ellsworth wasn't what he expected. Ellsworth told Colt to hold off applying for his U.S. patent—critical as that would be in protecting the potential fortune Colt stood to make with his revolver. Because of international patent laws, Ellsworth advised Colt to secure his patents in Europe before filing in the U.S. At the time, a prior U.S. filing would bar Colt from obtaining a patent in Great Britain. Only a month later, on August 24, 1835, Colt set sail for England, and then to France.

Although Colt successfully obtained his international patents, he returned heavily in debt. The costs for European patents were staggering: $676.50 in England and $341 for France, money he most likely borrowed from his father. While the $30 fee in the U.S. was substantial, enough to build up a surplus of around $150,000 in the patent office coffers, it was miniscule in comparison to what applicants paid in Europe.

Foreign patents in hand, Colt turned his attention to the U.S., where Ellsworth quickly shepherded Colt's application through the patent office. It issued on February 25, 1836, without any examination for novelty. But Colt now had what he needed to continue his dream, for a patent would give him the credibility to raise desperately needed capital. Colt's immediate plan was to raise funds to build a manufacturing factory in New Jersey. He first turned to his father for a short term loan. Colt's father was reluctant to advance any more money, preferring that Colt find other investors. Then Colt sought the help of his cousin Dudley Selden to help negotiate a loan, still hoping his father would join the effort and help provide some financial backing. Hearing of his son's plans, Christopher Colt advised him to use Henry Ellsworth to provide letters of recommendation to the Army and Navy Departments. They, not him, should be the ones paying for the revolver's manufacture. Still, Colt continued borrowing from friends and relatives and raised enough money to charter the Patent Arms Manufacturing Company of Paterson, New Jersey, on March 5, 1836. He produced his first revolving rifle at the end of 1836, then began working on pistols. The following year he would produce nearly 1,000 arms.

But sales were almost nonexistent. So Colt, now following his father's advice, went back to Ellsworth for letters of recommendation. Although Ellsworth did provide the recommendations, Ellsworth had little time to be of any more assistance. Ellsworth understood the deficiencies with America's patent system and was determined to do something about it. His attention had turned from mundane organizational issues to the real problems facing America's patent system, including the lack of substantive examination of patents.

What Ellsworth accomplished in a single year was, and still remains, unprecedented. With Ellsworth at the helm, 1836 proved to be the most monumental year in patent office history. Ellsworth had a plan to make life better for every inventor, not just Colt.

The superficial logistical matters he quickly set in order—the general state of disarray of the patent office, the garaged models, the backlog of patents awaiting the President's signature, and missing patent numbers. But the real problem Ellsworth quickly learned was that America's patent laws didn't protect inventors—and they weren't going to protect Colt.

The ideas behind how to change the patent system came from an unlikely source: the patent office machinist, Charles Michael Keller. As an adolescent, Keller spent most of his time hanging out at the patent office with his father, who was then the patent office machinist. By the time Keller's father passed away, Keller was twenty-one and had already logged eight years at the patent office. It was natural for Keller to be appointed to take his father's stead as the patent office machinist, presumably to work with the patent models.

When Ellsworth took his position as head of the patent office, Keller was twenty-five and had four more years of patent office experience under his belt, some as a patent clerk. It was a lowly position—one that paid less than half of what other government clerks received. But Keller knew the patent system and Ellsworth was smart enough to listen to him. Keller put together a list of changes that he felt necessary to fix the patent office. Ellsworth summarized these in a letter to Congress. Foremost on the list was the need to actually examine applications to make sure the ideas they contained were new. Keller had seen the abuses where unscrupulous individuals would visit the patent office, study a particular model, then fraudulently claim the idea as their own in a newly filed patent application. These fraudsters would then fabricate evidence to show they'd invented the idea first. This wasn't a new phenomenon. This had been going on ever since the days of Whitney. It was quite a cottage industry. With the certified patent in hand, signed by the President, the copier would seek royalties from legitimate businesses. It has been estimated that this business took in a half million dollars per year.

To ensure that patents would be properly examined, Ellsworth proposed to hire patent clerks with scientific backgrounds, to pay them a decent salary, and to provide them with a library of scientific journals so that they could examine applications in view of what was already known. Keeping the model requirement intact was also key to creating a successful patent system. Finally, Ellsworth wanted to publish issued patents and use them to generate a library of patents that would serve as a knowledge base for examining future patents. Other ideas included eliminating the need for the president's signature, a step necessary to eliminate the three months it usually took to do this, and allowing foreigners to patent in the U.S. but only if they paid the same amount that their home countries charged—$600 in England and $200 in France and Austria.

Ellsworth's most pressing need was for a new patent office. He fretted over the inability to display the models and that the patent office, like so many other federal buildings, would catch fire. With a surplus of $150,000, Ellsworth thought it was time to put this money to good use. America was the only country with a model requirement and these scientific models should be prominently displayed in a grand American museum.

While Ellsworth was working with Keller on ideas for overhauling the patent system, Maine's newest senator, John Ruggles, arrived in Washington with a keen interest in the patent office. Ruggles was not only a lawyer and a judge, but also a passionate inventor with a new idea for a cog railroad. And he was adamant that his idea be protected. After arriving in Washington, the first thing Ruggles did was to visit the patent office. Upon discovering the tattered state of affairs, Ruggles was as shocked as was Ellsworth just a few months before.

Together, the two men discussed what could be done and Ellsworth's plans for reform. When Ruggles discovered the genius behind Ellsworth's suggested changes, he asked to meet the clerk. The two, statesman and patent office machinist, soon gained a mutual respect, and together hammered out a new patent statute. The document marked the end of Jefferson's forty-three-year experiment, one in which Jefferson, frustrated with a myriad of issues involved in running a patent system, quietly bowed out. Now, after forty years of trial and error, rife with fraud, it was time to start anew. Ironically, what emerged was a patent bill not too dissimilar from the original Patent Act of 1790. Once again, patent applications would be examined for novelty, just as Keller—and Jefferson in 1813—had recommended. But this time, it wouldn't be by the Secretary of State. It would be by trained patent examiners. The new bill also provided $1,500 to purchase a library of scientific books to aid examiners, and it paid the patent office clerks a competitive salary. It eliminated the need for the President's signature and allowed foreigners to patent—as long as they paid the outrageous fees charged by their own countries. Applicants were still required to supply a model, which was to be publicly displayed. And for this to happen, the new bill asked for the existing $150,000 surplus to be used to construct a new patent office—a fireproof home to show off the nation's patent models. In what would prove to be an ill-advised decision, the act also allowed patent holders to apply for patent term extensions. If fourteen years wasn't enough, the patent holder could ask to extend his monopoly another seven years.

The statute crafted by Ruggles and Keller was signed into law on July 4, 1836, ten years after the passing of Jefferson and Adams, and sixty years after the signing of the Declaration of Independence. The signing date wasn't coincidental. Many considered this document to be America's second Declaration, not one to declare political independence from Britain, but to emphatically state America economic independence. With it, America gained a monumental advantage over Europe, not only in promoting innovation, but also in supporting America's groundswell of innovative energy. The Patent Act of 1836 would provide the framework to let inventors invent and investors to invest in their technology. It gave everyone the monetary incentive needed to put America's creative energy to work.

And it was only fitting to have the second Declaration crafted by a young, innovative patent clerk who, working on the front lines, saw exactly what America needed. He'd seen the anxious inventors, toting their treasures into Blodgett's Hotel. He understood their fears in putting their precious inventions into the hands of a bureaucratic system. And most important, Keller knew what drove them to invent. And with this understanding, Keller knew what it took to guide them along. In a sense, Keller's contribution to protecting innovation was similar to what Jefferson did in protecting freedom. Appropriately, Keller was the man to discover and draft America's second Declaration. To their credit, Ellsworth and Ruggles were smart enough to follow his lead.

After passage of the Patent Act of 1836, Ellsworth's title changed. He was now the first Commissioner of Patents. Not counting Jefferson, Keller was promoted to become America's first official patent examiner. His job was to compare the ideas found in patent applications with what was already known, in public use or on sale without the applicant's consent. In other words, the invention had to be new. The self-educated Keller took on this task, single-handedly examining every application in every field of technology. It was no small task. His first challenge was to whittle down the backlog of nearly 100 unexamined applications. By November, he would need to examine another 308 that had been filed under the new law. Over time, more examiners were hired, and in May 1845 Keller decided to study law and left the patent office—but not patent law. Keller became a successful patent lawyer in Washington, then moved his practice to New York.

With the new patent act in place, Ellsworth's organization skills were again put to good use. He began a new numbering system, with Ruggles' railway patent being assigned patent number one. Then he turned his attention to constructing the new patent office.

It was to be built on the block bordered by F and G Streets and Seventh and Ninth. Originally, this plot was reserved by city planners for a national church or pantheon. After the Rotunda took shape, this plan was abandoned. Yet a famous building would soon rest on the forgotten site, and it would become one of the nation's most visited. In only a few decades, the new patent office and model museum would receive more than 100,000 visitors a year who would see, as Abraham Lincoln put it, the real fuel of America's economy.

Everything now seemed in place for innovation in America to thrive. It had a new patent statute and a new patent office under construction. America's inventors were set—able to get good patents, then secure the investments they needed to bring their ideas to market.

But before that could all happen, the patent office would need to burn nearly to the ground, an omen signaling that America's patent system was ready to begin anew. On December 13, 1836, one of the post office clerks woke to suffocating smoke. The clerk rushed to wake the night watchman, and a quick examination of the building showed smoke billowing out of the southeast end. Still in his night clothes, the clerk frantically scampered down the street, raising the alarm. The ruckus awoke the neighbors, who rushed to assist. Both Ellsworth and Ruggles were eventually notified and rapidly made their way to the flaming hotel. Ellsworth first tried the front door, but found it blocked. He then tried another entrance but the smoke was so thick he was unable to enter. Meanwhile, Ruggles went next door to the fire house with a crew of volunteers to lug out an engine. But when he went to pump water, Ruggles discovered that the leather hose had disintegrated. Not a drop of water made it to the fire. In desperation Ruggles started a bucket brigade, but the laboring men stood no chance. By the time another fire engine arrived, flames were shooting out the windows. It was all over.

Only twenty-two years before, William Thornton had successfully fended off British soldiers from burning Blodgett's hotel to the ground, saving its precious models. Now, while the blueprints from a new fireproof patent office were still being drawn, these sacred relics were going up in smoke. The patents, the models, the drawings, the temporary home of Congress—all gone. In total, 10,000 patents were destroyed.

Ruggles immediately went to work to restore the documents, drafting a bill that requested patent holders to return their original patents so that clerks could make a certified copy. To motivate the inventors, any burned patent was declared invalid until a replacement copy was made by the patent office. Only 2,845 would eventually be restored.

And the new home for the patent office? Initially, it was moved into Ellsworth's home on C Street. It would take four more years for the new stone building to be completed.

However painful, death by fire turned out to be a blessing. It served to accelerate construction of the new patent office and to give inventors a fresh zeal toward the new patent laws. Although filings initially declined, after the 1836 Patent Act was firmly in place and the new patent office was completed, patents began to issue in record numbers. The reaction to the 1836 Patent Act can be shown by the following table:

Decade	Number of Patents
1837-1846	5,019
1847-1856	12,572
1857-1866	60,094
1867-1876	130,240

The number of issued patents can be compared to the population to determine the real rate of invention during the mid-nineteenth century, as follows:

Census Year	Population	Patents	Ratio
1840	17,069,453	473	1 to 36,088
1850	23,191,876	993	1 to 23,308
1860	31,443,321	4,778	1 to 6,525
1870	38,558,371	13,333	1 to 2,894

Put another way, the rate of innovation as determined from the number of patents increased six times from 1840 to 1850, nine times from 1850 to 1860, and 13 times from 1860 to 1870, as compared to the increase in population.

Why did this happen? It can only be attributed to the new patent system.

For the next two decades, consider the patents that issued: the revolver, the sewing machine, the reaper, vulcanized rubber, and the telegraph, to name just a few. With a flood of active patents, though, litigation was sure to follow. The ensuing patent battles would attract the best legal and political talent of our nation's history, —a sure sign that the protection of technology was critical to America's success.

Chapter Eight

THE PATENT OFFICE RESCUES COLT AND MORSE

Henry Ellsworth was determined to have his new patent scheme succeed. Americans needed faith in the new patent laws—enough to spur them on to invent. The proof wouldn't necessarily be whether more patents were granted, but whether inventors' ideas could be protected. And the best evidence to show that was for inventors to make money—lots of money. Nobody can argue with money.

Ellsworth didn't wait around to see whether America's inventors would be financially rewarded for their ideas. He left nothing to chance. Almost single-handedly, he delicately guided many of America's greatest inventors on the path to wealth and notoriety. Whether working secretly behind the scenes to grant patents, raising capital in U.S. and foreign markets, or constructing assembly lines, Ellsworth made sure his inventors would not only be awarded patents, but that the ideas behind those patents would be wildly successful in the marketplace. Stopping infringers—the key to protecting investment—wouldn't matter unless, or until, Ellsworth could get investors to have enough faith to begin investing in the first place. And investors would put up money once they saw patented ideas turned into commercially successful products.

The logistics—quick examinations, quality patent searches, completion of the patent office—happened almost as a matter of course. Like Jefferson, Ellsworth had skills that went well beyond the menial task of running a patent office. Even with the devastating financial crisis of 1837, a year in which fifty percent of the nation's banks failed, drying up capital markets and threatening reductions in patent office revenues, Ellsworth forged ahead undaunted. That year, Ellsworth increased the patent office staff from six to thirty-five. Although the staff increased, the number of granted patents did not. Initially, they dropped. In 1835 there were 757 patents granted, but only 435 in 1837. Some of this can be explained because a third of applications were now being rejected for lacking novelty—Ellsworth's quality assurance agenda. The other reason was the gripping depression. It affected everything, including construction of the new patent office, which dragged on until 1840, leaving patent examiners without permanent offices. And, with no storage, models had to be stored in the City Hall, further disrupting patent office efficiencies.

But as the economy loosened, patents began issuing at an unprecedented rate. In 1839, 800 applications were filed and 400 were rejected. In 1844, the number allowed reached 502. Ellsworth's vision was catching on.

But he still needed a success story—a poster child to validate his vision. Ellsworth needed a groundbreaking technology that would become a household name. If he could accomplish this, Ellsworth could claim that success was due to America's patent system, a system that would both invite future capital investment and also protect against copiers. To do this, Ellsworth's meddling in the nation's affairs went well beyond the reaches of the patent office and America's new Patent Act.

Ellsworth latched onto two inventions that could bring national attention to his platform: the revolver and the telegraph. His reasons for picking these two inventions went beyond the merits of the technology, though in their own right they each had the power to transform not only America, but the entire world. Ellsworth was probably influenced in his decision by who invented these technologies: two close friends of the Ellsworth family, Sam Colt and Samuel Morse.

———

Ellsworth's connection with the Colts and with Samuel Morse helped nurture the success of the telegraph and made Colt America's first patented millionaire. Without Ellsworth's influence, the names of Colt and Morse might now be unknown, striking from history the Colt revolver and the Western Union company.

In personality, Sam and Samuel were vastly different—rugged frontiersman versus eclectic professor. But in inventive drive, they were nearly identical. To launch their inventions, nothing was beneath them. Colt's antics began with his laughing gas tour to fund his fledgling munitions company. For Morse, it was a trip to the Capitol, visiting with legislators and stringing wires between Congressional offices. These two inventors were willing to try almost anything to market their inventions.

And both wanted fame, though for different reasons. Ellsworth would help them become two of the most popular men in America. And though the revolver and the telegraph had essentially nothing in common technologically, somehow Ellsworth labored behind the scenes so that Sam and Samuel synergistically propelled each other to success.

Ellsworth first began to form the idea to use Colt as a success story on July 24, 1835, when Colt appeared on his doorstep. Colt was ready to file his patent application and needed advice from the newly appointed head of the patent office. Fortuitously, Colt's father was good friends with Ellsworth and arranged the meeting. The advice Colt received wasn't what Colt or his father expected. Ellsworth explained that there were no international patent treaties, and if Colt filed his application in the U.S. before filing in England, he would eliminate any chance of patenting his revolver in Europe. If was a far cry from today's patent laws where most countries give a U.S. inventor 12 months to file for foreign protection after filing the initial U.S. application. Like most novice inventors of his time, Colt felt that the bulk of his business would come from Europe. That left him no choice but to borrow tens of thousands of dollars from his father, then speed to Europe to secure his patent rights overseas before starting his U.S. business.

Colt promptly packed his bags and by the end of August 1835 was bound for Liverpool. With his adventurous blood, being on the open seas again must have been refreshing, as if he really was going to make a name for himself. It would be one of many trips to Europe.

By October, Colt had his English patent in hand. He then turned to France, where he secured his second patent. Although quickly issued, the total cost for both patents was over $1,000. It was no wonder that few patents were granted in Europe during the nineteenth century. The fees were ten to twenty times more than in America. Colt's ingenuity in raising money with ventures like his laughing gas tour and borrowing from his father gave him this chance. Few other Americans could fund this large of an investment.

Colt returned to America just as Ellsworth was shepherding through Congress the Patent Act of 1836. It was good news for Colt. If the new bill before Congress was sound, the litigation fiasco that occurred with the Whitney cotton gin would never be repeated. Patents could now be enforced.

The future was promising. The economy was booming, and with his international patents secured, funding was all but a given.

Colt set to work to finalize his U.S. patent. Again he consulted with Ellsworth, who took Colt's $30 filing fee and shepherded the application through the patent office. On February 25, 1836, Colt was awarded his U.S. Patent No. 138 (later renumbered to 9430X after the patent office fire).

The power wielded by Colt's patent was all that Ellsworth dreamed it would be. With it, Colt persuaded his cousin Dudley Selden and several wealthy New York inventors to raise a whopping $230,000 and form the Patent Arms Manufacturing Company. Colt negotiated the rights to purchase a third of the shares, a $1,000 a year salary, and an expense account large enough to pacify any duke. Though still in its infancy, the U.S. patent system enabled Colt to convince businessmen to front the investments needed to begin manufacturing operations. The thought of investing in a monopoly on the revolver was just too irresistible to turn down.

Colt also had a little luck on his side. If he'd waited one more year, when the capital markets dried up during the panic of 1837, funding might not have been available. Several years later, when revolver sales failed to materialize, Colt would lose everything. It would take Morse and the Mexican war to save him again, resurrecting the revolver from the grave.

Boom times notwithstanding, the use of a patent in 1836 to raise a quarter million dollars was astounding, and exactly what Ellsworth was hoping to achieve. His new patent system gave investors the confidence needed to invest their money, with the assurance that they would have an exclusive market. And with ongoing boundary disputes with both European powers and the native Americans always threatening conflict, a healthy return on their investment in Colt's revolver seemed like a sure thing.

But money by itself doesn't mean success, especially in the hands of Colt. In a marketing free-for-all, Colt went to work wining and dining Washington's social elite, hoping to land a large government contract. All of Colt's efforts were extravagant, a trait Colt would exhibit the rest of his life. From expensive liquor to finely tailored suits, Colt found countless ways to use up his investors' money. Yet he failed to produce the orders. It was a hard lesson that all inventors eventually come to grips with: Although a patent might be able to entice investments, it does little to generate sales.

Still, Colt plowed ahead, armed with letters of recommendation from Ellsworth that he produced for the Army. Colt thought his big break would happen when he was invited to demonstrate his five-shooter at the Army and Navy trials at West Point in June 1837. The break didn't come. The Army Ordinance Department thought Colt's revolver was too complicated for battle situations. They'd stick with the tried and true single shot rifle.

In 1837 Colt did land one small order to supply revolvers to the army in its engagements in Florida during the Second Seminole War, with "Old Rough and Ready," Zachary Taylor, leading the charge. But even though the firearms proved a success in battle, further orders were minimal. Colt's revolutionary revolver just wasn't catching on. Even the revolver given to President Andrew Jackson received nothing more than a note of thanks. But that didn't stop Colt from spending his investor's money. The good life was just too good.

Colt eventually spent himself out of business. Marketing firearms in a down economy was fruitless, forcing Colt to consider the possibility of bankruptcy. Patent Arms was quickly losing money and sales were nonexistent. In only a few years, Colt's glory days were over. Lack of sales coupled with poor management drove the company into failure in 1841. Nearly a quarter million dollars was gone.

Colt's dark days quickly turned darker. When Colt had returned home from securing his European patents in 1835, he also brought home a dirty little secret—a pregnant mistress named Caroline Henshaw. Ever the finagler, Colt convinced his brother, John, to marry Caroline for a speculated $500. The evidence of Colt's bastard son came to light in September 1841 when John murdered his printer, Sam Adams, in an argument over a discrepancy in a bill. To hide the evidence of the murder, John hacked the body to pieces and stuffed it into a crate. But when the body eventually decomposed, John was discovered and arrested. He was tried for the murder, convicted, and sentenced to hang for the crime.

On the day of execution, November 18, 1842, John took his own life by plunging a knife into his heart, making Caroline Henshaw a widow. Colt felt an obligation to take care of her and her son, Samuel Caldwell

Colt. He did this by shipping them both off to Germany where Colt paid for her until she remarried. Colt eventually brought his son back to America and made him part of the family business. But, Colt's finagling wasn't enough to keep the whole affair from coming to light, tarnishing his already ailing reputation.

At age 28, Colt's life was a mess. His revolver appeared destined to become a relic, and many believed that he'd squandered all of his investors' money. Add to that his brother's death and life looked bleak.

Colt all but gave up on his revolver and turned his attention to his boyhood passion: blowing things up. Colt began experimenting with a waterproof bomb that he covered in tin foil. Essentially, the explosive device was waterproofed gun powder that Colt would ignite by using an electric spark from an underground wire. His new idea in hand, Colt wrote to President John Tyler bragging of a way to stop enemy ships from entering U.S. harbors, the precursor to later defense plans such as the MX missile or Star Wars. His plan was to submerge bombs in America's harbors, then remotely detonate them when enemy ships naively entered. Somehow, he was able to talk his way into a $6,000 grant to test his idea, then went to work to prove his idea.

Colt arranged for the use of a lab at New York University, sponsored by the Navy Department. Fortuitously, that was also where Morse, the ex-painter, now penny pinching, wild-idea professor, was spending his days tinkering with his telegraph. Behind all of this sat Henry Ellsworth, head of the patent office. Ever the matchmaker, he had found a way to link the two entrepreneurs together.

Ellsworth had known Morse since their days together at Yale. While Colt was dealing with his arms business, Ellsworth had been encouraging Morse with his telegraph. Now, the two struggling inventors were paired together.

Initially, Morse had little interest in Colt's antics, but eventually he realized they had a common problem. For Colt, it was sending an electric spark underwater to set off his charge. For Morse, he needed a way to convince Congress to give him a grant of $30,000 to run an underground line from Washington to Baltimore. His first attempt had fallen flat, with charges of self dealing, the result of taking on Representative F. O. J. Smith of the U.S. House as his business partner. Smith's bill to appropriate $30,000 was not well received when Smith's interest in the venture was revealed. Even a demonstration of the telegraph to President Martin Van Buren and his cabinet a month before didn't save Morse's cause.

Both Colt and Morse needed a waterproof electrical conductor to move forward—Colt for his explosives, and Morse for his telegraph. The magic turned out to be a copper wire covered with tarred thread.

To prove their concepts, in October 1842, Colt and Morse set up a two-day demonstration in New York harbor. Colt had strewn wire to an old ship in the harbor and blew it up in front of 40,000 spectators. The explosion was so spectacular that the Herald reported the ship went up in 1,756,901 pieces. Still, that wasn't good enough to sell Congress on the idea.

Then it was Morse's turn to show off his telegraph. Morse's lines ran from Manhattan to Governors Island, about a mile. Morse's initial attempt failed. His battery was too weak. Colt, with his underwater batteries, came to the rescue. He offered them to Morse, who hooked them up to his telegraph and successfully transmitted signals to Governor's Island, at least until a passing ship severed the cable.

Even with a successful test, Morse was losing hope of launching the telegraph. He was sliding down the same slope as Colt. Morse needed to demonstrate not only his product, but a demand for it. Without a long distance network of cables, that was impossible. After the panic of 1837, private investment was nonexistent and his request for funding was failing to gain the attention of Congress. Without some miraculous intervention, the telegraph too appeared doomed.

Though they were both in the same predicament, Morse and Colt had come to it by different routes. Colt was fortunate enough to finish his revolver before the financial crash of 1837 and had obtained funding, only to lose it. Morse found himself with a finished invention but no available capital. And while Colt was finding creative ways to spend a quarter of a million dollars, Morse had been frugal, teaching art in New York while working on his telegraph on nights and weekends.

It took until May 1838 before Morse had a proven concept. The new "Morse code" soon followed. And that gave Morse the desire to fund a company. Just like Colt, Morse wanted investment money—and lots of it. He petitioned Congress, then Martin Van Buren and his cabinet, all to no avail. Stymied in the U.S., Morse turned to Europe. If Congress wouldn't give him the $30,000 grant, another government certainly would. Morse was hoping that, like Colt, he could secure his patents in Europe, then use them as leverage to secure funding. Ellsworth tried to tell Morse not to bother with Europe—it was too late for any European patents. Morse had delayed too long and would now be prevented from obtaining European patents. By now, the secret of the telegraph was out. European scientists had entered the telegraph race and been publishing their ideas in the technical journals of the day. According to European patent law, these articles could prevent Morse from securing any European patents. Ellsworth told this fact to Morse and urged Morse to stay in Washington and negotiate with foreign ministers to take a license to the technology.

Against Ellsworth's advice, Morse went to Europe anyway—risking everything, as Morse himself put it. Facing too many roadblocks in America, Morse ventured his remaining money on getting patents in Europe. Europeans, he surmised, would appreciate and adopt his vision of long-distance communication. Morse went to work preparing prototypes to bring to Europe, all while working on his patent drawings for his U.S. application—"a most arduous and tedious process," as he put it. Before leaving, Morse asked Ellsworth to delay examination of his U.S. application until he returned, a last-ditch effort to protect any chance of getting a European patent.

Like Ellsworth predicted, European patent offices were hostile to Morse's requests. Britain wouldn't grant him a patent and no foreign governments were willing to join with Morse in a venture to install telegraph wires in their countries. The amount of infrastructure needed to run the wires did not strike these governments as a wise use of public funds. As a consolation prize, France awarded Morse his only patent in Europe.

For Morse, life was about creating. While in Europe, Morse met Louis Daguerre and learned about his new way to capture images. When Morse returned home with little hope of succeeding with the telegraph, he began dabbling with the Daguerre's technology and was the first American to make daguerreotypes in the U.S. By 1840, Morse had opened a daguerreotype studio where he trained Matthew Brady, the famous Civil War photographer, in using the technology.

Morse's new passion probably served to dull the pain caused by the failed telegraph business. The prospects of building commercial telegraph lines were bleak. Morse was so discouraged that he didn't even bother requesting to have his patent application examined. That all changed when Charles Wheatstone, a noted English scientist and inventor in the fields of electricity and spectroscopy, turned his attention to the telegraph. Wheatstone's U.S. patent filing and successful transmission over thirteen miles on lines tracking the Great Western Railway in England caught Morse off guard and propelled him into action. Morse complained to Ellsworth about Wheatstone's patent application, arguing that Wheatstone was a latecomer and should not be awarded a patent. Ellsworth reminded Morse that his own application was still on hold—a practice that was somewhat suspect and in essence served to extend the life of Morse's patent—and suggested that Morse reinstate his application. Later, this opened Morse's patent to attacks of nepotism and fraud. Morse followed Ellsworth's advice and formally asked Ellsworth to put the application into the examination queue. When it did get examined, the patent office wanted corrected drawings and for Morse to sign the oath. Just a few months later, on June 20, 1840 Morse's famous telegraph patent issued as U.S. patent number 1,647.

Even with his U.S. patent, Morse was unable to secure funding. Not only was it a blow to Morse, but also to Ellsworth who was hoping to prove the value of U.S. patents. With Colt's failure, Ellsworth desperately needed the telegraph to boom.

Raising capital was proving to be a formidable task. The infrastructure needed to make the telegraph succeed was too much for any single group of investors, and the government, still financially strapped, was unwilling to take the risk. Few fully understood how the telegraph could allow instant world-wide communication. Even for those that did, the thought of running wires around the country—and the world—was simply too daunting.

Yet Ellsworth encouraged Morse to keep trying. Ellsworth was politically connected, and a Congressional grant was still a possibility, though a remote one. In early 1843, Morse, backed by Ellsworth, reintroduced his request for $30,000. To help his cause, Morse went to the Capitol and strung wires from room to room. For some members of Congress, his antics only confirmed their suspicions that Morse was a frenzied professor and that investing in Morse's telegraph would be a waste of public funds.

As the sessions wore on, Morse's chances decreased. By March 3, 1843, Morse figured his best chance for passage of the bill had fallen through. It was the last day of the winter session and Morse, sitting in the visitor's gallery with Ellsworth, watched as the evening lights were lit. If the telegraph bill didn't pass in the next few hours, it would be shelved. An aide brought Morse the bad news—there were 119 bills ahead of his and that he had no chance this session. Frustrated, Morse gave up and went home to pack his bags. But not Ellsworth. This was like leaving a baseball game in the bottom of the ninth inning. As long as one out remained, there was a chance of winning. Ellsworth feverishly lobbied members of Congress late into the night. At the last possible minute, Ellsworth managed to get the bill approved, ahead of 117 others.

The next morning, Ellsworth's seventeen-year-old daughter, Annie, asked if she could personally deliver the good news to Morse. When she arrived at his boarding house, Morse was eating breakfast. Not only had Congress passed the bill, she told him, but the president had signed it into law just before midnight. Morse now had $30,000.

Elated with the news, Morse promised Annie that she could send the first message when the line was completed. With his new funds, Morse went to work preparing the lines. In Washington, he stayed with Ellsworth and stored his telegraph materials in the basement of the patent office.

A year later, Morse made good on his promise to Annie. On May 24, 1844, Morse sent the first telegraph from the old Supreme Court chambers of the Capitol to his business partner, Alfred Vail, in Baltimore. Annie had chosen a passage from chapter 23 of Numbers: "What hath God wrought!"

The Washington-Baltimore line was also working in time for the Democratic national convention, when James Polk was nominated for the presidency. For the first time, the telegraph spread the news of his victory back to Washington. In June 1844, Morse opened the line to the public— another way for Colt to advertise his wares. A message was sent that Colt had detonated a fuse of gunpowder in the Supreme Court building.

Motivated by the successful demonstration, Colt, with Morse's blessing, decided to start his own telegraph business named the Offing Magnetic Telegraph Association. In 1846 Morse had Colt run a line from New York to Coney Island to notify crews of ship arrivals. But as with his first foray into the telegraph business, Colt's telegraph company also fell flat, mostly because of mismanagement. That should have ruined Colt. But it didn't. War with Mexico was just around the corner.

Colt's old five shooters, the ones he'd shipped to Florida in 1837, had gained a reputation. The soldiers fortunate enough to have them swore to their effectiveness. When the Mexican War broke out, both Captain Samuel H. Walker and Major General Zachary Taylor wanted more, and they wanted them from Colt. But Walker wanted a better gun—a six shooter. He and Colt worked out a design and Colt was awarded the contract for 1,000 revolvers. The problem was that Colt didn't know how to make a six shooter. Nor did he have a factory anymore. But through Ellsworth, Colt gained a contact with Eli Whitney, Jr. who had taken over his father's gun-making operations. The Whitney arms manufacturing business had been well established since the turn of the nineteenth century. Whitney had turned to gun making when Southerners refused to license his patent on the cotton gin and he needed another source of income. Ellsworth's connection to the Whitney family went back to the cotton gin days when Ellsworth's father-in-law collaborated with Eli Whitney during Whitney's venture with the cotton gin.

By May 1847 Whitney had completed the revolvers. They arrived in the Mexican territories four days before Walker was killed. But it was enough to put Colt on the path to becoming the first patented millionaire. Morse would follow, but only Morse would make it to the Rotunda. Not as the artist of a portrait, but as its subject. It was all thanks to a good idea, to Ellsworth, his patents, and some luck.

Chapter Nine

McCormick and Goodyear Secure Their Rotunda Fame

Morse wouldn't be the only inventor represented in the Rotunda. Because of the success of the reaper and vulcanized rubber, so also would McCormick and Goodyear.

In 1837, still unsure whether he could make a go of his reaper, McCormick tried to diversify his business interests. He turned to the iron furnace business, then became a victim of the great panic. With no other prospects for success, McCormick returned to his reaper, giving his full attention to catching up with Obed Hussey, who by now had proven he could sell his own, barely-workable reaper for a handsome profit. In 1840, McCormick announced his entry into the reaper game in Virginia.

The problem was that McCormick still didn't have a reaper that he liked. And if he didn't like it, neither would the farmers. So McCormick kept toying with his design. And while he did, Hussey pressed ahead with his own reaper—with essentially no competition. McCormick was furious. Not only had Hussey beaten McCormick to the patent office, he'd now beaten him in the market as well. By 1841 Hussey was selling in five states. McCormick now needed not only a functioning reaper, but one that was superior to Hussey's.

And that is just what he did. Still, that didn't make McCormick the inventor of the reaper, even though history has generally portrayed it that way. The truth is that McCormick was no more the inventor of the reaper than Isaac Singer was of the sewing machine. It was a collaborative effort by multiple inventors, each making his own improvements. By borrowing each other's advances, they eventually developed a commercially workable machine. What McCormick excelled at was inventing the improvements that made the reaper commercially attractive to farmers. With improvements such as reciprocating blades shielded by metal fingers, a reel to bring the grains to the blades, a divider to isolate the grain, and a platform to hold the man gathering the cut grain, McCormick had a decided edge.

In 1844 McCormick, satisfied with his new and improved reaper, came up with the idea of challenging Hussey—his only real competitor— to a reaper's duel. The competition turned out to be one-sided, with McCormick cutting seventeen acres to Hussey's two. More important than the win was the publicity it generated. American farmers began to see the unbelievable efficiencies made possible by the reaper.

McCormick bolstered his market position when he recognized that the reaper was ill suited for the hilly Virginia terrain. He turned west, opening a new factory in Chicago in 1847, nearer to the flat, open prairie where the reaper excelled. At the time, Chicago had a population of just under ten thousand, and was thought of as a run-down cow town. To finance his operation, McCormick joined with the mayor of Chicago, William B. Ogden, who put up $25,000 for fifty percent ownership. They built a factory on the banks of the Chicago River.

Sales, however, were sluggish, not so much because other competitors were taking his market share, but because there simply was no market. McCormick had to convince farmers this was their future. McCormick well understood that having a great product, no matter how good the patent protection, does not generate sales by itself.

For that, McCormick invented a new business strategy: selling on credit. To advertise, McCormick set up harvesting competitions, pitting his reaper against anyone willing to take him on. McCormick's reaper started gaining the attention of farmers, especially after a farmer named James Hite cut 175 acres in a week and exclaimed, "My reaper has more than paid for itself in one harvest." McCormick used this to his advantage and signed up Hite and other farmers as his sales force.

McCormick also promoted sales by offering a written guarantee. With $30 down, McCormick collected another $90 only if the machine could cut a certain amount of grain in an hour. He marketed hard, taking out large advertisements with testimonials. Added to all this, McCormick bound his agents by exclusive contracts with exclusive territories.

Over time, his marketing and business strategies began paying off. During 1848, his first year in his new Chicago factory, McCormick produced a meager 100 reapers. That number quintupled to 500 by the next season. In 1851, McCormick was busy manufacturing 1,000 reapers, a number that would escalate to 23,000 by 1857, turning a handsome profit of $1.25 million.

Unfortunately for McCormick, when the first machines rolled off the line in 1848, his original patent had expired and Congress refused to grant an extension. As McCormick became successful, so did the competition, riding on McCormick's success. Other competitors began testing the market and later proved to be much more astute than Hussey. In 1848 McCormick found at least 30 rivals in the field, a figure that would continue to escalate.

Fortunately, McCormick was a savvy businessman. He realized that he could not rest on his original patent rights because they had expired before the market had ever developed. Instead, he filed additional patents with each improvement, including two filed in 1845 and 1853, forming a legal iron fence around his reaper. These improvement patents turned out to be more valuable than his first because they covered concepts that made the reaper a practical farming implement. And those "improvement" patents would be the ones McCormick would assert against his competitors.

———

Goodyear had his own rags-to-riches story. After years of poverty and persistence, in 1837 Goodyear thought he had the solution to the rubber problem. He took out his first patent covering the use of nitric acid to process rubber. Although this process produced the best rubber so far, it still wasn't commercially feasible—as Goodyear discovered when 150 mailbags ordered by the government melted to nothing.

Still, with his 1837 patent Goodyear found a backer in William Ballard that allowed him to move into a factory on Staten Island, this time bringing his family with him. But Goodyear wasn't immune to the effects of the crash of 1837. Although Goodyear didn't have any money to lose, his potential investors did. Forthcoming funds abruptly ended. Somehow, word of Goodyear's experiments reached William Ely, who invested $10,000. But Goodyear ran out of money again, forcing his family in 1839 to sell off their possessions and landing him in debtor's prison. At his lowest point, Goodyear received $50, barely enough to continue his work in Springfield, Massachusetts. It was here where Goodyear had his breakthrough. As legend has it, during a heated argument, Goodyear waved a rubber sample in his fist. It spurted out and landed on a heated pot-belly stove. When he scraped away the charred material it was dry and springy. Goodyear's discovery was that rubber "carelessly brought into contact with the hot stove" turned into leather. It was the miracle cure for rubber, the magical treatment that would keep it from melting during the summer and becoming brittle in winter. A kind of pasteurization, but for rubber. It was the spark of an idea that led him to vulcanization—named after Vulcan, the Roman god of beneficial fire, who also made his way onto the Rotunda—and a newfound zeal.

Even with his prospects brightening, it took Goodyear until 1843 to work out his vulcanization process. He lodged his patent application the same year. The recipe called for 25 parts rubber, 5 parts sulfur, 7 parts lead, and turpentine heated to 270 degrees. His core patent issued June 15, 1844. That same day, Goodyear's nemesis, Horace Day, was also busy at work attempting to obtain his own vulcanization patent.

The bad blood between Goodyear and Day had started years before. Although Goodyear knew his vulcanization process required high temperatures years before filing his patent application, he needed a commercial-sized furnace that could maintain these temperatures. How to maintain a precise, constant temperature to conduct his experiments proved to be a monumental problem. As a point of honor, Goodyear didn't want to file a patent application until he was sure he had a commercial product. And it was the issue of the furnace that led to his chance meeting with Day, one that would haunt Goodyear for decades to come.

In 1841 Goodyear redesigned his oven so that it could reliably maintain elevated temperatures. It was an expensive proposition, but one that ultimately led to his next discovery: that for vulcanization, the temperature needed to be between 245 to 300 degrees F. The magic number turned out to be right at 270 degrees. To help pay for this redesign, Goodyear managed to convince the owner of the adjacent business, Horace Cutler, to pitch in $300. In return, the two men formed a joint venture to make rubber shoes. Cutler was evidently unaware that previous Goodyear investors had yet to receive any measurable returns on their investments. Still, Cutler put in his money, and when the first batches of shoes came out "half baked," Cutler wanted out, and he wanted his money back. Of course, Goodyear didn't have any money, so he did what he always did: He stalled, hoping to buy time and avoid another trip to debtor's prison. His absent-mindedness to matters of finance plagued him his entire life, robbing him of a life of luxury. But to Goodyear, it was never about the money, only the invention. Goodyear's altruism was no consolation for Cutler. As a last-ditch effort to recuperate his losses, Cutler sold the shoes,

95 of them, for $26 to a Mr. Horace Day of Brunswick, N. J. And that's where Goodyear's problems really began.

If Goodyear had known anything about Horace Day, he would have never let go of those ninety-five shoes. Day's scandals included leaving his pregnant wife and five-year-old child to marry his own cousin. His *Times* obituary had nothing good to say about him, blasting his contentious litigation with Goodyear, the indisputable inventor of vulcanized rubber, his belief in Spiritualism where he claimed to talk with the dead, and his far-fetched businesses that gobbled up his rubber fortune, such as one that proposed to lay a pipeline four feet in diameter from Niagara Falls to New York to supply the city with pressurized air.

Goodyear had nothing of Day's extravagance. He had the appearance of a grown-up Dickens' character: slim build, clean shaven, coal black hair, and innocent eyes. And unlike Day and his escapades, Goodyear loved hard work and doting on his large family. Even so, as Day would discover, Goodyear had plenty of fight in him, too.

After receiving Cutler's shipment of shoes, Day immediately struck up a correspondence and convinced Cutler to move to New Jersey to help him set up a competing shop, though Day may not have revealed his true intensions. They agreed on a salary of $30 per month and Cutler headed south. But after only a few months, Cutler became disillusioned with Day and his brash personality and wanted to return home. Day was furious when Cutler approached him with the news. Day was desperate to understand what Goodyear had discovered. Stay long enough to tell me Goodyear's secrets, he demanded. Cutler finally capitulated to Day's unrelenting demands and offered to divulge the factory secrets for $75. Day was appalled at the amount and bargained the price down to $50, and for that trivial sum, Day hacked his way into one of the greatest secrets of the nineteenth century. But even with the recipe, Day had problems making vulcanized rubber. The process didn't seem to fall into place as Cutler had explained. In desperation, Day journeyed to Springfield incognito to visit Goodyear's plant. The two met and Goodyear refused to let Day—or anyone else—inside the factory.

And while Day fretted about his own failures, life wasn't getting much better for Goodyear. After redesigning his oven he was again out of money—taking another one of his many trips to debtor's prison. After his release, Goodyear plodded ahead on a shoestring budget. Perfectionist until the end, Goodyear wasn't yet ready to tell the world what he had discovered, and he refused to file his patent application until he was confident that every aspect of his vulcanization process was complete. From a legal perspective, it was a poor decision. Goodyear waited until 1844 to file his application, and the delay cost him his largest market. Across the ocean, England recognized patent rights based on the first person to file the application—not the first to invent the idea. Knowing this, Thomas Hancock, who reverse-engineered some of Goodyear's rubber samples, filed his own application in England before Goodyear, essentially stealing the English market.

Meanwhile, Day, who still had Goodyear's recipe, tried the same trick in the States and filed his own patent application. The idea was to try to get his own patent on somebody else's invention, as in the days of Whitney. But Day's application was brought to the attention of Henry Ellsworth, who recognized the scheme. It was just this type of fraud he'd set out to stop with the Patent Act of 1836. On June 15, 1844, the same day Goodyear's patent issued, U.S. patent commissioner Henry Ellsworth rejected Day's application, citing Goodyear's application, which had been filed the previous January. "Your application for Letters Patent, for an alleged improvement in preparing India rubber, has been examined, and rejected for want of novelty. An application for a patent for a similar process to yours was made on January last, by Charles Goodyear, of New York."

That didn't stop Day's shenanigans. He simply starting knocking off Goodyear's shirred rubber goods, regardless of who held the patent. Shirred rubber goods, including stockings, caps and gloves, constituted a substantial market. These goods were made by gluing a previously stretched, thin rubber strip between two pieces of cloth, then allowing the strips to contract (shirring the cloth). In retaliation, Goodyear sued Day in both 1844 and 1845. As Singer would do to Howe in the sewing machine wars, Day worked to manipulate public opinion against Goodyear, taking out slanderous newspaper advertisements. Day had the audacity to claim that Goodyear wasn't really the inventor of vulcanized rubber, claiming that credit should go to a man named Hayward, one of Goodyear's research assistants. Goodyear responded in kind, publishing a statement that he could produce any number of witnesses to contradict Day. Then Goodyear took one final jab. Because he was too busy inventing and didn't have time to monitor the scoundrel's daily attacks, he claimed, he concluded, "I cannot expect to look after his notorious bulletins *Day* after *Day*."

The rhetoric continued for the next year. In an about-face, right before the trail began in November, 1846, Day agreed to settle. This time, Day published another notice that he and Goodyear had settled their differences and that Day now had a license to produce shirred goods under Goodyear's patent. Day continued with his publications, offering a $50 reward for anyone who would identify an infringer.

What Goodyear would soon learn is that Day never had any intention of paying Goodyear under the license, even for the mere pittance of three cents per yard. Moreover, Day wasn't about to limit himself to producing only shirred goods. Anything with rubber it in was fair game. Who cared about Goodyear's patent?

Day's next foray was into rubber shoes, which he marketed as "Day's Best Patent Japan Rubber Shoes," a clear belittlement of Goodyear's India rubber shoes. The two were headed for another showdown.

Chapter Ten

SINGER STARTS
THE PATENT WARS

1851 was a watershed year for innovation in so many ways. It was the year of the Great Exhibit in London, perhaps the greatest of all the world's fairs. Morse, Goodyear, McCormick, and Colt were all in attendance. So was the sewing machine—although missing its biggest proponent, Isaac Singer, who was busy in America starting the nation's patent wars.

The 1850s proved to be the decade of the most prolific patent litigation—or any litigation—in America's history. Lincoln himself was involved, as well as his three most prominent cabinet members: Chase, Seward and Stanton. So too was the Secretary of State, Daniel Webster. At the heart of it all was the battle over the sewing machine.

As the effects of the 1837 financial downturn loosened, more inventors began to invent and file patent applications on those inventions. It started with the rubber boom, then extended to munitions, agriculture machinery, new materials, and telecommunications. Industries began to burgeon with the increase in capital investments. A prime example was the manufacture of sewing machines.

Constructing a viable sewing machine had proven to be fascinatingly complex. The race to invent the sewing machine began long before anyone had ever heard of Elias Howe or Isaac Singer. It involved the combination of dozens of ideas thought up over several decades. Perhaps the most vexing problem was how to pull the thread up from the bottom side of the fabric. Early attempts focused on mimicking the motion of a human hand: A needle was pushed completely through the fabric, then forced back through the other side.

To his credit, Howe was the first person to look at the problem from another point of view. Instead of trying to copy the motion of manual stitching, Howe conceived of using an eye-pointed needle in combination with a shuttle (which holds the bobbin), which brought up a second thread from the opposite side of the fabric to create a lockstitch. This eliminated the need for the needle to pass entirely through the cloth. This same technology is still found in today's sewing machines and was the long-awaited breakthrough that reenergized inventive activity on the sewing machine.

To apply for a patent, Howe created a patent model that many consider to be the finest model ever submitted. Howe's patent application issued on September 10, 1846. While multiple times faster than sewing by hand (at 250 stitches per minute), Howe's sewing machine had problems, the foremost being that he used a curved needle that tended to break. Without a robust design, Howe's initial attempts at commercialization failed, his sewing machines languishing on store shelves. Frustrated, Howe left for England to sell his idea to British tailors. Not surprisingly, the English also shied away from Howe's machine, finding the design still too crude.

While Howe was overseas peddling his design, the race to perfect the sewing machine in America was off to a brisk start, with everyone building upon Howe's lockstitch technology. Dozens of inventors feverishly went to work, each making his own incremental improvements. Nearly all these ideas were deemed patent worthy, resulting not only in technologies that would eventually make the sewing machine commercially practical, but in an arsenal of patents. It was the beginning of a patent arms race that would inevitably play itself out in America's courtrooms.

The sewing machine technology gelled around 1850, with multiple companies having commercially workable designs. But the person who came up with the best design was Isaac Singer, who replaced Howe's curved needle with a straight one that hung vertically. Singer rose to fame when, using his last dollar, he fixed the floundering Blodgett sewing machine, one of the many designs incorporating Howe's original lockstitch concept. Singer did this by moving the needle vertically and adding a foot pedal. This allowed the fabric to remain horizontal during stitching. Singer fixed other problems with Howe's design, like altering the shuttle and varying the tension on the thread. Like Howe, Singer too received a patent, which issued on August 12, 1851. Singer's machine—the Jenny Lind—was a class above anything else, producing 900 stitches per minute, an estimated 25 times faster than what could be done by hand.

Singer's sewing machines caught the attention of the garment industry, inviting dozens, if not hundreds, of sewing machine startups into the field. Although Singer's machine spoke for itself, Singer would later prove to be a marketing genius, rolling out a highly effective national advertising campaign that helped make the Jenny Lind the clear market leader.

Singer also had a tremendous ego. His arrogance in thinking he was above the law, both with his company and with his personal life, landed him in court on numerous occasions. What Singer didn't understand, or more likely, what he simply chose to ignore, was that Howe had patented the lockstitch, and no court would ever care that Singer also had a patent on a design that cured the deficiencies in Howe's sewing machine. A basic tenant of patent law is that having your own patent doesn't get you around an earlier patent. To complicate matters, dozens of others had also improved Howe's original design and they too had patents on components of the sewing machine. Everyone in the sewing machine industry was infringing somebody else's patent. Singer probably figured that Howe couldn't possibly stop them all, and besides, other people were infringing his patent and he wasn't throwing a fit. What was the big fuss about patents?

The melee that ensued set the pattern for the patent assertions during our own computer/semiconductor revolution. The situation of overlapping patent rights is unavoidable. Every technology needs improvements and along with those improvements come additional patents. This was especially so with the sewing machine, where hundreds of tinkerers were needed to create the final design. The result was that Howe wasn't the only one holding a patent on the sewing machine technology.

But a blocking patent, like Howe's patent on the lockstitch, can be used to keep everyone out of the market, unless the patent holder is adequately compensated, usually by paying a royalty. And this was Howe's strategy—to collect royalties from every sewing machine using his lockstitch design, rather than selling his own sewing machines. Because of Howe's strategy, new sewing machine companies risked getting sued for patent infringement even if they received an improvement patent on Howe's original design.

Many believe that these kinds of blocking patents hinder competition. However, if patents aren't awarded on the main principle, nobody wants to invent. The result is often industry-wide patent litigation.

To Singer, Howe's original lockstitch patent along with the dozens of other improvement patents awarded to other sewing machine companies were no roadblocks. Perhaps speed bumps at best. Singer, with his fiery temperament, was determined to get his way. Singer attempted to dominate the sewing machine industry, disregarding any legal threats from Howe. After all, Howe wasn't even selling sewing machines. At best, all Howe had was a piece of paper from the patent office.

Howe came to disdain Singer late in 1850 when Howe, recently returned from England, was strolling down Broadway in New York City. When he reached the storefront window of Smith and Conant's, an early department store, he saw a demonstration of a lockstitch sewing machine that started his heart racing. For the past year, Howe had been obsessed with the myriad of new sewing machines that were coming onto the market, all using his patented design. While his own venture to make a commercially workable sewing machine had failed, his efforts in obtaining a patent had not. Now backed by a financial speculator, Howe was ready to start "suing the infringers of his patent for royalties."

Directly in front of him as if to challenge his recently sworn oath to "sue the infringers" was none other than a freshly minted Jenny Lind sewing machine from I.M. Singer & Co. With resolve, Howe tracked down Singer in his machine shop, ready for a confrontation. There he found an immense man with a full beard and waxed moustache that shot skyward like a sprouting bean plant. As Howe would soon discover, Singer's temper was as fiery as his red hair.

Howe didn't waste time getting down to business. He demanded that Singer pay him a royalty of $2,000—the wrong thing to ask a man who was not only out of cash, but who had such an unwieldy temperament that he later would flee to Europe after severely beating one of his wives (he had several families on the side) and his daughter. In a rage, Singer threatened to kick Howe down the steps of his machine shop. It would take another year before Howe mustered up enough courage to face Singer again.

By 1851 the sewing machine industry was poised to take off, and Howe knew he had to act or it would be too late. Facing risky and expensive litigation, Howe again approached the brazen man, this time demanding Singer pay him $25,000 to settle their differences. Why Howe thought Singer would pay $25,000 when only a year before he had refused a $2,000 settlement offer is a mystery. Howe should have known that the trip would be a waste. His demands were ignored. The Singer establishment viewed Howe as nothing more than a "perfect humbug," one that "never invented anything of value."

Out of patience, Howe exercised his only remaining option. He sued Singer, along with a handful of other sewing machine manufacturers in Boston, and the case came to trial in June 1852. It was the spark that would set off a series of explosions.

Howe's timing in suing Singer was perfect. Singer was in desperate need for cash to finance the manufacture of the Jenny Lind. Initially sales were slow, mostly because of a skeptical public who had seen previous machines not live up to their claims. When Singer started his venture, he had looked to George Zieber for financial help. Zieber, a publisher who had once financed Singer in developing an alternative printer's type, agreed to provide modest financing for a fifty percent interest in Singer's patent. With Howe's lawsuit, however, Singer needed more money than Zieber could provide.

But nothing was beneath Singer. Ignoring Zieber's share in his patent, Singer took on another partner, his patent attorney Edward Clark. Singer offered Clark Zieber's fifty percent interest in exchange for free legal services. Then, Singer went looking for a way to swindle Zieber out of the share that he'd just given to Clark. The opportunity came only a few months later when Zieber was seriously ill. Singer approached his bed and broke the bad news to Zieber—that Zieber was dying of an incurable disease. It was a flat-out lie, but it was all Singer needed to convince Zieber to sell his interest for a mere $6,000.

Howe's first sewing machine trial began in June 1852 in the Circuit Court of Massachusetts. It lasted three weeks. What Singer and the rest of the sewing machine industry underestimated was the strength of Howe's patent. As Singer would later be forced to acknowledge, Howe really was the first one to invent the lockstitch. Because every sewing machine used this technology, infringement was almost a forgone conclusion. By July 12, 1852, Howe had obtained preliminary injunctions against the accused infringers. That was enough for most of the manufacturers to fall in line, paying Howe an exorbitant royalty of $25 per machine.

But Singer was another matter. He refused to bow to Howe. The battle between them moved from the courtroom to the editorial pages of the *New York Daily Tribune*. On July 29, 1853 Howe wrote a warning:

The Sewing Machine—It has been recently decided by the United States Court that Elias Howe, Jr., of No. 305

Broadway, was the originator of the Sewing Machines now extensively used. Call at his office and see ... a certified copy, from the records of the United States Court, of the injunction against Singer's machine.... You that want sewing machines, be cautious how you purchase them of others than him or those licensed under him, else the law will compel you to pay twice over.

On the same page, Singer presented his own case. "For the last two years Elias Howe, Jr., of Massachusetts, has been threatening suits and injunctions against all the world who make, use or sell Sewing Machines." Singer then claimed that he was selling his machines rapidly and that the "public do not acknowledge Mr. Howe's pretensions, and for the best reasons." He discounted Howe's machines, saying that they didn't work and that Howe wasn't the original inventor. If they wanted the best "Sewing Machine," they needed to come to Singer.

This prompted a charge of libel from Howe, who sued the *New York Daily Tribune*. In turn, *Scientific America* immediately condemned Howe's rhetoric, as did a host of other papers.

Singer, who often struggled to maintain self-control, was now livid. As Singer told a friend, he would have killed Howe given the opportunity by squashing his neck with his foot. And Howe wasn't the only beneficiary of Singer's wrath. Those who signed up for licenses from Howe immediately became Singer's enemies—robbers and damn scoundrels, as he referred to them.

Singer knew that he was running out of options. Both the sewing machine industry and the courts took the position that Singer's Jenny Lind sewing machine infringed Howe's patent. He desperately needed another strategy, but Singer was never without a scheme. There would be more battles between Singer and that "robber."

Singer still refused to pay homage to Howe. With all of his energy focused on fighting Howe, his company suffered, but that only served to intensify his resolve. The war was raging on every front—the press, the courts, and now with Singer's next strategy, it would return to the patent office. Because everyone knew that Singer infringed Howe's patent, Singer needed another approach to free himself from Howe's threats.

In late 1853, Singer had his plan. He would show that Howe didn't really invent the sewing machine. If he could do this, all his troubles would go away. Without delay he scoured the world for prior art, hoping to find evidence of an obscure sewing machine that would invalidate Howe's patent. Singer searched everywhere, including European patent offices. He even conducted an investigation into whether a machine had been built in China. When those leads failed, Singer resorted to more shrewd tactics by concocting one final scheme to invalidate Howe's patent.

From his searches, Singer found a man named Walter Hunt, a prolific inventor with twenty patents to his name. In 1834 Hunt had supposedly built a sewing machine with an eye-point needle and shuttle. Singer's plan was for Hunt to file a patent application with the patent office—a full eighteen years after Hunt allegedly built his sewing machine—then claim Hunt should be awarded Howe's patent. If all went according to plan, the patent office would award Hunt the patent on the lockstitch, rather than Howe. This was based on a development in U.S. patent law that awarded a patent to the first person to invent the idea, not the first person to file his patent application. Fabricating evidence of earlier designs was not unusual. This same tactic was also used against both Goodyear and Colt when they tried to enforce their patents.

In 1853, Singer carried out his plan, paying Hunt—along with a team of re-constructionists—to reassemble an eighteen-year-old pile of rusty parts into some semblance of a sewing machine. Then, in the fall of 1853, he paid for Hunt to file a patent application and ask the patent office to have Hunt declared the original inventor of the sewing machine. Taking matters to the extreme, Singer also funded an advertisement in the *New York Tribune* where Hunt declared himself—not Howe—as the original inventor of the sewing machine. Hunt continued by bragging that he was pursuing his claim with the patent office.

These "rusty claims" provoked the ire of the *Scientific American*, who denounced the unscrupulous tactics. While the patent office was reviewing the matter, Howe persevered in the courts, promptly suing several Singer retailers in Massachusetts Circuit Court for an injunction. The suit against the retailers was filed in February of 1854 and would be decided by August of the same year. As summarized by the court:

> This was an application for a provisional injunction to restrain the defendants from infringing the letters patent [No. 4,750] granted to Elias Howe, Jr., September 10, 1846, for an improved sewing machine, by the use and sale of the Singer machine, so called. The defendants denied the novelty of the invention of Howe, and relied, in support of their denial, mainly upon an alleged invention of Walter Hunt, in 1834.

Because Singer's infringement was a foregone conclusion, his only chance of succeeding was to show that Howe's patent wasn't valid—using the reconstructed model of Walter Hunt. The outcome would depend almost entirely on whether Singer could prove that Howe wasn't the first to invent the sewing machine. Not only did Singer latch onto the remains of Walter Hunt's old machine, but Singer also presented a new replica of Hunt's 1834 machine.

The court didn't take kindly to Singer's charade. "The first inventor [Hunt] gave nothing to the public. His so-called invention was only an idea, never carried out in a machine that could anticipate one subsequently invented."

Having dismissed Singer's invalidity arguments, the judge quickly wrapped up the case. "The other question, as to the infringement, remains. … The weight of testimony … is strongly preponderating in favor of the plaintiffs…. The result is, that the plaintiff's patent is valid, and the defendants' machine is an infringement. An injunction is granted."

This was the end of the line for Singer, especially after the patent office in February 1855, after hundreds of pages of dubious testimony, also concluded that Hunt's claims were "rusty" and declared that the sewing machine patent rightfully belonged to Howe. Singer reluctantly capitulated and paid Howe $15,000 plus a royalty of $25 per machine like everyone else. Although the royalty was significant, Singer could still pay that amount and stay in business, especially if everyone else had to pay the same royalty. In a rather surprising move, together the two men took out an ad in *Scientific American* warning that if any remaining sewing machine makers didn't take a license, they too would be sued for infringement.

But with hundreds of improvement patents on the sewing machine, this threat only highlighted the looming infringement problems facing the sewing machine industry. By 1867 there were around 900 patents on the sewing machine. The sewing machine industry wasn't alone in seeing a boom of patents. The same would happen with the reaper.

Almost overnight dozens of new suits were filed across the country. The litigation was so fierce and widespread that the press came to the conclusion that the sewing machine companies would sue themselves out of existence. Singer, still paying royalties to Howe, again found himself a target, being named in at least twenty suits filed in four different courts scattered along the East coast. Not wanting to miss out, Singer also sued on his own patent, taking on Grover & Baker and Wheeler, Wilson & Co. in Philadelphia. Even Howe found himself named in some of the suits.

Like today, the number of documents required to be produced took the attention of these companies, leaving them little time to peddle their wares. Tens of thousands of pages of testimony resulted from all the trials. Though a pittance compared to today, it was still time-consuming when it all had to be recorded on a typewriter.

The end result was that the lawsuits crippled the market to the extent it almost didn't exist. Howe was collecting only a few hundreds dollars a year on his patent. Although there were certainly other reasons for the slow sales, including the need to break social norms where women were discouraged from working with machines, the all-out patent wars were mostly to blame.

It was like the arms race between the Soviet Union and the United States, but this time, somebody had pulled the trigger. Unless the whole industry came to its senses, it would cease to exist. For some, this all foretold the failure of the patent system.

Then, Orlando B. Potter, president and legal counsel of one of the defendants, Grover & Baker, came up with an idea to use the market economy to fix the problem. Because Henry Ellsworth had done his job in creating a patent office that generated valid patents, these legal instruments were valuable assets, with a calculable market value. In other words, since the patents were enforceable, they could also be bartered, just like any other asset.

Potter's idea was to pool the patents, with everyone in the pool chipping in a license fee. The basic idea was for each person in the pool (referred to as a "combination") to grant a license to everyone else in the pool. In addition, the members had to pay a royalty on each sewing machine sold. This money was divided according to who owned the most valuable patents—with the "core" patents being owned by Howe, Singer, Wheeler, Wilson & Co., and Grover & Baker. The collected royalties were paid out to Howe and the "core" patent holders, with the remaining funds being used to cover the costs of administration as well as to enforce the patents against infringers who did not join the combination.

At first, Howe resisted because he had the most to lose. He eventually gave in, especially after being offered $5 for each machine sold in the U.S. and $1 abroad. The provision that convinced Howe to join the combination was when Potter guaranteed the combination would have at least twenty-four licensees, essentially guaranteeing Howe a healthy revenue stream.

With that, the Great Sewing Machine Combination was formed, and members paid $15 for each machine sold. While each member received a license to all the patents in the pool, they were free to charge their customers any price. Howe did receive his $5 per machine sold, although in 1860 this was reduced to $1 per machine sold when the membership price charged by the Combination was reduced from $15 to $7. Although Singer grumbled over the deal—as he always did when he had to give up his money—the $15 was barely noticeable as Singer sold his machines for $125, while the cost to make a machine was only $23.

The resulting royalty arrangement made Howe a millionaire. In 1860 his profits had reached $444,000. Then Congress granted him a seven-year patent extension based on Howe's petition claiming that he was not justly compensated for his patent during the original fourteen year term, allowing him to collect two million dollars by the time his patent expired—all without his ever successfully producing a sewing machine.

But the beauty of the Combination was that it almost overnight settled all the disputes, allowing the sewing machine industry to explode. And, it put faith back in the patent system. Americans understood that they could protect their ideas, and if there were conflicting patents, solutions could be fashioned to make everybody a winner.

Still, in starting his patent battles, Howe taught a generation of inventors how to sue. We have never looked back. With no commercially acceptable product, Howe was paving the way for today's non-practicing entities (NPEs) or "patent trolls": inventors who have no viable product yet secure valuable patent rights used to block others from coming into the market.

Chapter Eleven

COLT TURNS TO THE COURTS

After struggling for twenty years, Colt had—with the help of the Mexican War—turned himself into a millionaire almost overnight. But he was facing a major problem. By 1848, copycats were hitting the gun stores and Colt's patent was about to expire. Valid for 14 years, the patent was set to lapse in 1850. Without his patent, Colt would be hard pressed to stop gun makers from producing revolvers.

Fortunately, Colt had one of the country's best patent attorneys, Edward N. Dickerson, who knew how to make the most of the patent system. Seeing that litigation was imminent, in 1848 Dickerson filed a request to have Colt's patent "reissued." This was a process where Colt could argue that the patent was partly defective; that is, Colt had left out some kind of improvement or failed to fully explain his idea. Because the patent was defective, Colt could fix the mistake in the patent office. It was a common practice, done by Morse as well, and it enabled Colt to strengthen his patent before initiating a lawsuit.

A year later, Dickerson petitioned to have Colt's patent extended for another seven years. To qualify, Colt had to show that this was a valuable patent to the country and that Colt was unable to reap the expected profits. As part of this process, Colt had to give notice of his intention to let others, usually competitors, object to the extension. When all the evidence was considered, Congress granted the petition, extending Colt's patent until February 25, 1857.

With more patent life, Colt focused his attention on wielding his patent to stop the infringements. In the spring of 1851, Colt decided to enforce his patent against Massachusetts Arms in a Boston court, which also was where Goodyear was preparing to assert his patent against Day.

Boston was a good choice for Colt to start his fight. It was close to home, meaning he would have more time to spend on other pursuits, such as spending his new-found wealth. Colt had recently purchased a speculative tract of land on the Connecticut River and was studying how to keep out the flood waters so he could build a factory. At the same time, he was also making plans to sail for England where he would exhibit five hundred guns at London's Crystal Palace Exposition with a not-so-novel marketing strategy—pouring free brandy down the parched throats of his visitors.

The trouble with Colt's most formidable infringer began when two of Colt's former employees started manufacturing their own revolvers in Springfield, Massachusetts. The company was aptly named the Massachusetts Arms Company. Secretly, Colt had his cousin purchase one of the infringing revolvers from the factory. "I do not want them to know the arms are for me," he told his cousin. "I want them immediately & you will confer a grate favour on me by getting & forwarding them as soon as you can, & much oblige."

Convinced the revolvers infringed his patent, Colt had his patent lawyer, Dickerson, promptly file a patent infringement suit in Boston in May 1851. The case was assigned to Justice Woodbury and would turn out to be one of America's liveliest and most publicized patent infringement suits. Colt's choice of lawyers proved to be a wise one. During the next three years, four of America's greatest patent cases would come to trial, and Dickerson would have his hand in three of them. It was Dickerson who a year later would assist Daniel Webster in the Great India Rubber case, then four years later would advise William Seward in McCormick's reaper case. The notoriety generated by these cases would attract not only Dickerson, but also many of the most influential figures of the nineteenth century. Never in America's history have so many prominent individuals fought over such significant, ground-breaking technologies, ones that would drastically change the country's future: the revolver, the telegraph, vulcanized rubber, the sewing machine, and the reaper.

With Colt still in England, Dickerson moved ahead with Colt's case against Massachusetts Arms, generally making all of the strategic decisions without input from Colt. If he had waited for Colt to return, the case would have been over. Unlike today's trials that can last two or more years, the case was tried and decided by August of the same year, while Colt was on his sales campaign in Europe.

What made Colt's case so interesting—besides the immense popularity of the revolver and Colt's propulsion into riches and stardom—were the charges of fraud and corruption raised by both sides. Foremost among these were raised by the infamous Rufus Choate, who represented Massachusetts Arms. His presence alone brought in hoards of press into Boston, giving Dickerson all he could handle.

Rufus Choate was the Johnny Cochran of his day, and perhaps the leading trial lawyer of the decade. But this was far from an O.J. murder trial. Choate had already had his fair share of those. His most famous was a case in which he defended a Boston man who had allegedly sneaked into his lover's room and slashed the prostitute's throat with a razor. Then the man attempted to burn the whorehouse to the ground. Choate spun several tales as he crafted his defense, including the argument that the woman had committed suicide. If it wasn't that, her death must have been the result of the defendant's sleepwalking, where in his sleep he had sauntered across town and killed his lover. Whatever the reason, the jury seemed captivated by Choate's charisma and acquitted the man.

For Massachusetts Arms, Choate was the ideal selection. If anyone could invalidate Colt's patent, Choate could. And he planned to do just that.

Although the patent laws in America were still developing, they had come far enough so that patent rights were delineated by claims—a summary of what the inventor believed to be the inventive concept. Patents could have multiple claims, and to infringe the entire patent, the defendant had to infringe only one of them. "I am instructed to request you to consider [the claims] separately," Justice Woodbury told the jury in Colt's case. "If the defendants use one of them it is enough; and it is of no consequence to this result whether they use more than one, or more than two."

But to get to that point, the jury had to understand exactly what Colt had invented. "What did Colt do?" Justice Woodbury asked them. "He undertook, undoubtedly, from all that appears in the case, and from the specification, to get the power, through a revolver, of having more discharges in a short space of time than by a single barrel. That is one great essence of this principle of revolving fire-arms. He introduced revolvers, undoubtedly, which might be fired oftener within the same space of time."

Infringement of Colt's patent was almost a given, and so Choate turned to another tactic: to argue that Colt's patent was invalid. While an invalidity defense is a common strategy in any patent case, Massachusetts Arms went about this in a very unconventional way. In an attempt to confuse the jury, Choate argued that Colt wasn't the first to invent the revolver. It was a classic Choate tactic—move to another issue where it would be easier to fabricate evidence and confuse the jury. The only problem was that each of the guns Choate showed the jury were made after Colt invented his revolver.

None of these shenanigans slipped past Justice Woodbury. "The Smith gun," the judge said, "which is the one pressed most strongly as to date, was not finished, according to the mass of the testimony, until 1833— some considered that it was in 1834—but it was not finished, so as to be an operative piece, until 1833; and if so, it is wholly immaterial to go into any consideration as to how near it resembled the plaintiff's, for if it was of subsequent date, it does not impair or impeach his."

But the attempt to deceive the jury went far beyond irrelevant prior art. Choate had dug up evidence of an early revolver made by a Mr. Colburn. During the trial, it came to light that this revolver had been put together just for the trial. Dickerson told the jury a story about how the gun had been lugged half way around the country to rust it up and make it look worn so that it would look antiquated.

The one credible argument raised by the Massachusetts Arms defense involved the discovery of a gun described in an early French patent. If the defense could show this was the same thing Colt tried to patent, then the judge could rule that Colt's patent was invalid. As for "what is called the French or Coolidge gun," the judge said, "that was patented abroad about 1818, and published in 1825, so that it is early enough in date; and the only question is, whether the combination and the machinery used there, to effect the object, was the same in substance or principle with that in Mr. Colt's."

To this, the judge told the jury that the French pistol used a spring to revolve the chamber while Colt's gun operated by "drawing up the hammer, and in that way causing the chamber to revolve without any coil spring." His charge was for them to decide whether the two were based on the same principle. However, by the mere way in which he asked the question, it was clear how the jury would decide the issue, especially when the judge said that he "must confess my inability to do it." While not proper in today's courts, this interjection by a respected judge certainly nudged the jury to what appears to be a just ruling.

Choate also personally attacked Colt. His clients were furious about the argument Colt had used to obtain a patent term extension—from 14 years to 21 years—on the basis that Colt's patent wasn't sufficiently commercialized during the original term. The defense argued that Colt failed to give the proper public notice of his intent to apply for an extension by altering papers at the patent office. Knowing Colt, it's certainly likely that the allegation had some basis of truth. The allegation impressed the press, but not the judge. "Whatever may have been the cause, ... the commissioner made the extension; it is, therefore, in point of law, valid, and I see in it no evidence of fraud. If there was any fraud at all, it would be in the commissioner, rather than anybody else. The jury cannot find fraud without evidence."

In August 1851 the jury concluded their deliberations and found Colt's patent both valid and infringed. As was customary, the damages issue went undecided, leaving the parties to resolve the issues among themselves. However, with an injunction in place, Massachusetts Arms would need to settle with Colt or risk losing the ability to license Colt's patent in the future. This allowed Dickerson to work out a deal with Massachusetts Arms to pay Colt $15,000.

With all the controversy surrounding the trial, Colt decided to have the entire trial transcript published. But that didn't end the matter. Colt's tussle with Massachusetts Arms was just beginning.

While Colt was in Europe celebrating his victory, Massachusetts Arms was busy manufacturing more infringing revolvers without obtaining a license. With Colt still overseas, Dickerson took it upon himself to file a second lawsuit in October, 1852, against company officials Hiram Terry and Edwin Leavitt. This time, however, the case was filed in New York. As with the Boston press, the New York papers were in heaven, latching onto the same stories of intrigue and deception—Colt bribing patent officials to extend his patent, and the defendants fabricating evidence about prior art revolvers.

At trial, Dickerson used his previous script, taking issue with the defense for doctoring up old weapons in an attempt to show that Colt wasn't the first to invent the revolver. He began with the Smith gun.

Now your Honors will see (showing the gun to the Court) ... that in the slot was once the cock of this gun, and that through that screw hole was the centre of that cock. Where are they gone? Now the lock is changed, and another sort of lock occupies its place—one in which the manner draws back straight and strikes on end. This is obvious from inspection, and bears the most convincing evidence of fraud on its face.

As if that weren't convincing enough, Dickerson then told the court that "a little domestic vinegar" was used to rust the metal and make it old.

Then he turned to the Colburn gun.

Here is the 'original sin,' the very one which was made up for the Boston trial by the defendants there, and rusted with sulphuric acid and browning, to make it look old. It has been produced here, under the pressure of my notice, by the defendants, most reluctantly, and has been accompanied by the affidavit of T. W. Carter, of Springfield, to show that he did not rust the gun, nor know of the trick till after it was done.

After ridiculing each of the prior art guns in turn, Dickerson summed up his case. "Neither the 'Smith,' ,Colburn,' 'Fisher,' or 'Collier' gun has any means of unlocking and relocking the chamber with the lock, in the act of cocking, and this, I submit, entitles the plaintiff to the relief prayed for."

As in the first trial, Massachusetts Arms argued Colt used fraud in obtaining his patent extension. The court was in no mood to entertain this argument, telling Massachusetts Arms that it already had its day in court on the matter. The court then held in favor of Colt.

Choate had tried his best, but a win for his client with such unfavorable facts was unlikely. Even so, the dirty lawyering for which Choate was famous would also attract the attention of Horace Day, who engaged Choate to take on Daniel Webster in the Great India Rubber Case.

With two victories in hand, Dickerson went on the offensive. Wanting to head off any future lawsuits, Dickerson published a circular telling the industry that, "All rotary arms constructed with such combinations, whether made by the Springfield Arms Co., by Young & Leavitt, by Allen & Thurber, by Blunt & Syms, by Marstin & Sprague, by Bolen, or by any other person, are a plain violation of Col. Colt's patent." The notice concluded with a warning: "You will please to take notice to desist forthwith from the sale of any REPEATING FIRE ARMS, in which rotation, or locking and releasing, are produced by combining the breech with the lock ... except such as are made by Col. Colt, at Hartford. ... I shall proceed against you and hold you responsible for damages, if you persist in the sale of any such arms."

The notice, along with two successful lawsuits, had the desired effect. The industry took heed and the infringements virtually stopped. In fact, Colt's patent policing activities essentially ended until his patent expired in the beginning of 1857. After that, Colt had other patents on other aspects of the revolver that in 1859 took him back into the courtroom.

Only one company, Allen & Thurber, came forward to ask for a license. On December 29, 1852, they paid Colt's attorney $15,000 for permission to keep manufacturing revolvers. The one problem with this settlement was that Dickerson brokered the deal while Colt was in Europe, trying to stop the numerous knock-offs being sold there. When Colt returned, he was furious with Dickerson and wanted to undo the deal. Dickerson told Colt, in essence, to calm down and not push his luck, which had been better than that of any other American inventor. He told Colt to let the matter drop, and Colt reluctantly did.

Chapter Twelve

GOODYEAR SEEKS OUT DANIEL WEBSTER

While Elias Howe was fuming over the physical violence threatened against him by Singer, sixty miles south in Trenton, N.J., Charles Goodyear had also reached his limit. After sweating and toiling for sixteen years over tens of thousands of rubber samples, enduring visits to debtor's prison, and finally discovering the vulcanization secret almost by accident, success now appeared to be a sure thing—except for Horace Day's meddling.

With Day's bold move in stealing Goodyear's vulcanization process, then attempting to patent it as his own, the normally mild-mannered Goodyear was livid, taking this as a personal affront to his honor. By 1851 Goodyear was in a state of depression—the result of nearly a decade-long battle with Day. The issues between the two men had reached an impasse, and the issues between them were about to be decided in a Trenton, New Jersey courtroom. For both men, the ire between them was all consuming—hotter than fires they used to vulcanize the very rubber they were fighting over.

That same year, Goodyear had ventured to England, intent on breaking into European markets, even without a British patent. His best opportunity was at the Great Exhibition in London's Hyde Park, which opened on May 1, 1851. Goodyear wasn't the only American inventor showing off his wares. Sam Colt also had a booth inside the enormous Crystal Palace to tout his 500 revolvers. Cyrus McCormick paid a visit as well, with an array of reapers—as did his main rival Obed Hussey. When the two squared off in a reaper's duel, thousands of farmers watched McCormick demolish Hussey.

Goodyear's display outshone those of both Colt and McCormick. Spending $30,000, Goodyear made sure that even the walls of his exhibit space were made out of rubber. Goodyear worked tirelessly for months, suffering through a stint of gout, to make everything perfect, as he always did. Instead of a single room like most of the other exhibits, Goodyear rented three, all filled with an imaginative assortment of rubber goods. It was floor to ceiling rubber. When visitors entered the display, they were greeted by a rubber desk. After that came rubber balloons, canes, doll heads, and ship sails. The Goodyear exhibit was so popular that it was continuously thronged by visitors and praised by the London papers.

Goodyear's days of glory in England and at the Great Exhibition were short-lived. When he returned home, Goodyear was again out of cash, and now what appeared to be his final battle with Horace Day was imminent. Tense as the days leading up to trial were becoming, having his day in court would at least bring some measure of relief. What Goodyear needed was finality, an end to the years of feuding.

Horace Day had no conscience when it came to selling Goodyear's rubber. Not only had he failed to pay Goodyear his license fees, Day was now producing all sorts of rubber products. Goodyear desperately needed a victory.

Back in September, 1850, Goodyear had come to the conclusion that more litigation with Day was inevitable. By then, Day was openly hostile toward Goodyear's patent, claiming it was a fraud and a swindle. He even petitioned Congress to repeal Goodyear's patent. Day's petition was simply ignored.

Partly to deal with Day, the Goodyear Shoe Association was formed. This was one of the first trade associations in America and included shoe manufactures who had taken a license to the Goodyear patent. The association collected three cents per shoe to raise money to stop Day. The association took out advertisements warning consumers that any rubber shoes manufactured by Day infringed Goodyear's patents.

Bolstered by the backing of the trade association, Goodyear instructed his attorney to file suit against Day in Trenton, seeking an injunction. Probably because the case was to be heard on Day's own turf, in early 1851 Goodyear's attorney also filed a second suit against Day in Boston, then sought to stall the Trenton case so that Goodyear could get a sympathetic ear in Massachusetts.

But Goodyear needed another lawyer on his team—the best lawyer in America—to argue his case. With the help of the Goodyear Shoe Association, he considered his options. It would have been easier if money wasn't an issue—but it was. With the large expenditures in London, Goodyear's financial means were limited. Still, that didn't stop him when he realized that retaining Daniel Webster, the current Secretary of State to Millard Fillmore, was within the realm of possibility.

Goodyear's team offered Webster $10,000 to argue his case of patent infringement against Day, with a $5,000 kicker if he won. George Griswold, a New York rubber manufacturer, offered to throw in another $1,000. Even with his government responsibilities, Webster readily accepted.

In some ways, Webster and Goodyear were very similar. They were both passionate and acted on principle. They both were also horrible with their check books. While Goodyear spent his money on rubber, Webster spent his on the fineries of life—an estate in Marshfield, high-priced vintages imbibed while dining the country's social elite, and plenty of bad investments. They were both heavily in debt, and they both couldn't stand to see a wrong go unchecked. It was a perfect match.

With his other commitments, it was impossible for Webster to handle all the details of the trial. Instead, Webster was saved for the closing arguments, like a closing pitcher in the ninth inning. Attending to the details were a team of patent lawyers, including Dickerson who had just handled Colt's case.

Snow came early to Boston in the fall of 1851, essentially shutting down the city on October 28. So too was Goodyear's Boston case. It now appeared that it might not even come to trial that year. This delay was unusual. That same year, Colt's case had been filed, tried, and decided in a single summer.

While Daniel Webster was making his way from the nation's capital to Boston, Day's counsel, Rufus Choate, moved the Boston court to hold off hearing the case until after Goodyear's trial in Trenton. According to Day, the previous September Goodyear had forced him to prepare for the Trenton trial, then claimed at the last minute that he wasn't ready. Day viewed this as nothing more than a ruse to stay the Trenton case so that Goodyear could do battle in Boston, a much more favorable venue for Webster. Webster was now coming to Boston to ask the judge if the court could work around his busy schedule as Secretary of State and try the Boston case the day after Christmas.

Goodyear's request for a Christmas trial in Boston was denied on October 25, 1851, when Judge Sprague agreed with Day and stayed the Boston case. For Sprague, the decision was easy. It saved his Christmas holiday and let the media circus have their day in New Jersey. When the news reached New Brunswick, papers reported that the night air rang with celebration as Day's supporters cheered in the streets, waving banners and torches. They'd regained their home court advantage.

On March 23, 1852, Daniel Webster made his grandiose appearance in Trenton for Goodyear's long-awaited battle with Day. The case was reported by journalists all along the East coast. The initial problem facing the judges was the limited number of seats in the federal courthouse. There were so many who were unable to find seats that they simply milled about the courthouse, attracting food vendors hoping to cash in on the circus. Even the state legislature adjourned for those who wanted to attend.

Of itself, having Daniel Webster on his team didn't make a win for Goodyear a forgone conclusion. Goodyear's legal team still had to deal with Rufus Choate, who after the Colt cases fully appreciated what fabricated evidence would succeed and what wouldn't.

Famous as Choate was, the truth was that nobody really cared about what Day's attorney would say in his arguments. They were all waiting to hear from the last of America's greatest orators. In the days before football, baseball, rock concerts, and movies, the celebrities of the mid-nineteenth century were statesmen, explorers, orators, and inventors. In 1852, the greatest of these was Daniel Webster, and his agreement to try a patent case for Goodyear while still in office was front page news across the country. It was Webster who had eulogized Jefferson and Adams in 1826 with a two-hour oration in Boston. It was Webster who had negotiated the border treaty between Canada and the United States, and who had served as a U.S. Senator from Massachusetts, working beside Henry Clay and John C. Calhoun as a supporter of the Union cause. Webster's accomplishments also included arguing 223 cases before the Supreme Court, and three times running for the Presidency. And it was Webster who coined the phrase that America's government was "made for the people, made by the people, and answerable to the people" in perhaps the most eloquent speech ever delivered to Congress in his debate with Senator Robert Y. Hayne on protectionist tariffs. Now, at age 70, he had obtained a mythic status, akin to the revered Founding Fathers. Webster had spent time with Jefferson at his Monticello estate and had dined with James Madison. Now, only five months before he would pass away, Webster would give his last great oration.

The first defeat for Day was his request for a jury trial—the main reason he'd fought to have the trial in Trenton. The case was brought in equity, a legal proceeding based on English law that typically involves a remedy other than money damages. Two judges were assigned to hear the case. These judges had the discretion to decide the case on their own—something we desperately need today. Knowing of Choate's reputation, the judges likely had no desire for weeks of testimony and jury deliberations, particularly if Choate's was going to attempt to confuse a jury with bogus testimony about prior inventorship—a tactic he'd used several times before in the Colt trials.

After the first day of arguments it was clear that a new, larger venue was in order, so the trial moved to the Mercer County Courthouse with its 700 seats. For the next two days, the sides argued the merits of their cases. Proving infringement was unnecessary. Everyone knew Day infringed—and he essentially conceded this point. Instead, he tried to prove that Goodyear wasn't the first to invent the vulcanization process, just as he'd tried to argue six years before. Choate presented testimony of other inventors who claimed to have invented the same thing as early as 1833. Choate was audacious enough to suggest that Day himself had invented the process before Goodyear—a risky argument when many knew about Day's payoff to steal Goodyear's secret.

Webster remained silent, carefully taking notes as he crafted his closing arguments. He'd leave the mundane details of dismantling Day's witnesses to others on Goodyear's legal team. As part of their case, Goodyear's attorneys told of how Goodyear had invented vulcanization, how Day had paid a bribe to discover the secret, how Goodyear had previously beaten Day in litigation. They also showed how Day had failed to pay Goodyear on their 1846 settlement agreement where Goodyear allowed Day to produce shirred rubber goods, making Day wealthy why Goodyear fell deeper into debt. Goodyear's team concluded by showing Day's inconsistent position in arguing that Goodyear's patent was invalid when Day had previously taken out newspaper advertisements touting its validity.

By day four, the two seasoned lawyers were ready to present their closing arguments. Choate went first, taking a full five hours to wrap up his case. When he was finished, the local papers, clearly on Day's side, heralded the argument as the finest ever heard. The Day constituents were hopeful fate was on their side.

But Webster was still to come. "Tomorrow Webster commences," the *Times* reported, "for which he is fully prepared. The Court Room continues crowded." While many suppose that the breathtaking legal fee offered to Webster was why this great man, burdened with so many other responsibilities, decided to take on Goodyear's cause, Webster's opening statement tells the real reason. On May 14, 1852, the *New York Times* published Webster's argument.

"I believe that the man who sits at this table, Charles Goodyear, is to go down to posterity in the history of the arts in this country, in that great class of inventors, at the head of which stands Robert Fulton, in which class stand the names of Whitney, and of Morse, and in which class will stand 'non post longo intervallo,' the humble name of Charles Goodyear." Webster went on, highlighting Goodyear's travails and the circumstances that brought about his "great invention":

> Notwithstanding all the difficulties he encountered he went on. If there was reproach, he bore it. If poverty, he suffered under it; but he went on, and these people followed him from step to step, from 1834 to 1839, or until a later period, when his invention was completed, and then they opened their eyes with astonishment. They then saw that what they had been treating with ridicule, was sublime; that what they had made the subject of reproach, was the exercise of great inventive genius; that what they had laughed at was the perseverance of a man of talent with great perceptive faculties, with indomitable perseverance and intellect and had brought out a wonder as much to their astonishment, as if another sun had risen in the hemisphere above. He says of his cell in the debtor's jail, that 'it is as good a lodging as he may expect this side the grave;' he hopes his friends will come and see him on the subject of India rubber manufacture; and then he

speaks of his family and of his wife. He had but two objects, his family and his discovery. In all his distress, and in all his trials, his wife was willing to participate in his sufferings, and endure everything, and hope everything; she was willing to be poor; she was willing to go to prison, if it was necessary, when he went to prison; she was willing to share with him everything; and that was his only solace. ... Mr. Goodyear survived all this, and I am sure that he would go through the same suffering ten times again for the same consolation. He carried on his experiments perseveringly, and with success, and obtained a patent in 1844 for his great invention.

To add a personal touch, Webster told of his own experience with rubber, beginning with all of the foolish investors in the early 1830s—before Goodyear's invention. One of Webster's friends had sent him a coat and hat made of this new magic material. But when he left it on the porch one cold evening, the coat had turned stiff as a board. In the morning, his neighbors thought this was the new manor of the estate.

The crux of Day's argument had been that either Day himself or somebody other than Goodyear was the first inventor of vulcanized rubber. Webster's response to this assertion took on a flavor that was different from the moving oration in which he opened. It was one calculated to smoke out the fraud committed by Day when he stole Goodyear's idea, then claimed it as his own:

There is not a single question of fact in the case we have said, on which the court can feel the least doubt. We assert that Goodyear is the first man upon whose mind the idea ever flashed, or to whose intelligence the fact ever was disclosed, that by carrying heat to a certain height it would cease to render plastic the India rubber, and begin to harden and metallize it. If there is a man in the world who found out that fact before Goodyear, who is he? Where is he? On what continent does he live? Who has heard of him? What books treat of him? What man among all the men on earth has seen him, known him, or named him? Yet it is certain that this discovery has been made. It is certain that it exists. It is certain that it is now a matter of common knowledge all over the civilized world. It is certain that ten or twelve years ago it was not knowledge. It is certain that this curious result has grown into knowledge by somebody's discovery and invention. And who is that somebody? If Goodyear did not make this discovery, who did make it? Who did make it? If the other side had endeavoured to prove that some one other than Mr. Goodyear had made this discovery, that would have been fair. But they do not meet Goodyear's claim by setting up a distinct claim of any body else. They attempt to prove that Goodyear was not the inventor, by little shreds and patches of testimony. Here a little bit of sulphur, and there a little parcel of lead; here a little degree of heat, a little hotter than would warm a man's hands, and in which a man could live for ten minutes or a quarter of an hour; and yet they never came to the point. There are birds which fly in the air, seldom lighting, but often hovering. Now this is a question not to be hovered over, not to be brooded over, and not to be dealt with as an infinitesimal quantity of small things. It is a case calling for a manly admission and a manly defense. I ask again, if there is any body else than Goodyear who made this invention, who is he?

Webster continued, attacking Day as the only person on the earth who denied that Goodyear was the true inventor. Webster closed by reminding the judges that Day had already lost another patent infringement to Goodyear suit six years before, and what right did he now have to reargue the same case? Then he sat down. It would be the last argument the great Daniel Webster would ever make in a court of law.

Needing to conduct some urgent business, Webster left that evening for Washington, missing the celebration banquet thrown by Goodyear, now confident of his victory. The next month Goodyear returned to

England to continue his legal battles overseas. Webster didn't remain in Washington but left for his Marshfield estate, in desperate need for some relaxation. In May 1851 he and a friend ventured off to a local fishing hole. On the way, their carriage axle broke, tossing Webster headlong onto the ground. The resulting injuries to his arms and head were too much for him, already in failing health. By the end of summer, Webster was bedridden, still waiting for the Goodyear decision.

It came on September 28, 1852. Justice Grier rendered his decision in favor of Goodyear:

> And yet when genius and patient perseverance have at length succeeded, in spite of sneers and scoffs, in perfecting some valuable invention or discovery, how seldom is it followed by reward! Envy robs him of the honour, while speculators, swindlers, and pirates, rob him of the profits. Every unsuccessful experimenter who did, or did not, come very near making the discovery, now claims it. Every one who can invent an improvement, or vary its form, claims a right to pirate the original discovery. We need not summon Morse, or Blanchard, or Woodworth, to prove that this is the usual history of every great discovery or invention.

Justice Grier's decision mainly turned on the outcome of the previous lawsuit where Day had wholeheartedly taken a license and acknowledged the validity of Goodyear's patent. Day's spurious evidence of other inventors was more than frowned upon by the judges. They understood Day's motives. Grier concluded his decision by granting Goodyear a perpetual injunction as well as ordering an accounting of all of Day's infringing sales of shirred goods.

The decision temporarily revived Webster's spirits and he believed that he would overcome his injuries. He took to his correspondence, paying off old debts and catching up on business matters. His exuberance was short lived. Barely two weeks after the decision, Webster's health rapidly deteriorated and on October 24, 1852, he silently passed away. His funeral was held at his estate, where 10,000 paid their respects. Five times that number attended a memorial service in Boston. For Webster, the $15,000 fee turned out to be a godsend as it allowed him to pay off his nagging debts before his death.

The Great India Rubber trial is one from a bygone era—a battle between two legends, Choate against Webster. Both presented their arguments, not laced with technical jargon, but in simple layman's terms. Was Day making vulcanized rubber? Yes. Did Day invent vulcanization before Goodyear? No, because if he did, why would Day have previously taken a license? And the astronomical legal fee paid to Webster? A mere pittance compared to the millions we charge today. Why? Because one didn't need dozens of experts to opine on infringement—opinions that are never understood by juries, anyway. And, unlike today, there weren't thousands of possible defenses, or reasons why the infringer could infringe even if he was doing something different. Yet the outcome was as it should have been.

What if Goodyear's case were retried today? There is a high probability that Goodyear's patent would have been struck down, or Day could have found a way not to infringe. That seems unbelievable, but it is true. That's patent litigation in today's America.

Chapter Thirteen
THE LEGAL ELITE JOIN THE FIGHT: LINCOLN AND HIS FUTURE CABINET TAKE SIDES ON THE McCORMICK REAPER CASE

While sales of McCormick's reaper were initially slow, by 1854 it was clear that the reaper was a viable alternative to the scythe. It was also the year when the American public bought in to Singer's sewing machine. Between the two, they created a buzz of economic change in America. With all the fanfare surrounding the litigation involving the revolver, vulcanized rubber, and now the sewing machine, McCormick must have understood that success meant not only competition, but also inevitable patent litigation.

No longer was Hussey McCormick's main threat. Dozens of other reaper models now flooded the market. The brilliant marketing strategies of the past—the credit buying, the reaping competitions—none of that was enough to keep ahead. Like his peers, McCormick knew he had to turn to his patents. And he entered the litigation fray by suing William Seymour. He picked an easy fight, taking on a lightweight, a common strategy when enforcing patents against a industry full with infringers.

To argue the case, McCormick engaged one of the country's most prominent trial lawyers, William H. Seward, Lincoln's future Secretary of

State, who was stabbed several times in the head and throat on the night of Lincoln's assassination. Originally a general lawyer, Seward got hooked on patent cases, eventually encountering the likes of Daniel Webster. By the time of his engagement with McCormick, Seward was a seasoned patent litigator.

On October 24, 1854, Seward argued the case against Seymour in the Circuit Court in Albany, New York. He gave a stirring oration about how McCormick's reaper, harvesting grain at an acre an hour, turned the biblical verse "cursed be the ground for thy sake" into a blessing by eliminating the need for both the sickle and "the sweat of thy face." Seward continued by explaining the specific features that made McCormick's reaper so successful, and how these were the aspects of McCormick's patent that Seymour was infringing. Seward's arguments were successful, and McCormick was awarded $7,750 in damages.

But McCormick's foray in the world of patent litigation was just beginning. McCormick had more on his mind than Seymour, or even Obed Hussey. More formidable entrants were flooding into this promising new industry. On the marketing side, McCormick kept the pressure on with his reaping contests, hoping to beat his competitors on the strength of his reaper alone. But with so many competitive reapers, his luck began to turn.

Before, McCormick had toted his competitive edge in reaper contests, winning nearly all of them by large margins. But that changed during a reaper contest held on Wednesday, July 5, 1854, where McCormick's reaper was outflanked by one from John H. Manny of Rockford, Illinois. As reported by the *New York Times*, the competition included reapers from Manny, McCormick, Ketchum, Allen, and Atkins. Using hooded horses, the machines were run through high grasses, each hoping to cut the most. Manny emerged as the clear winner and was praised by the *Times*, who concluded their piece by stating that "The farmers are fast gathering their crops, and I do not know what they would do without mowers and reapers."

McCormick's ire with Manny continued to brew. While Manny's reaper handily won the contest, McCormick felt it did so only because it incorporated features covered by his patents. When Manny again beat McCormick in a contest at the Paris Exposition of 1855, McCormick had had enough and made Manny his prime target.

Fortunately, by 1856 McCormick was selling reapers at a rate of 4,000 reapers a year, and he now had the resources to take on Manny in the courtroom. The ensuing battle would last three years, including an appeal to the U.S. Supreme Court.

Made popular by Abraham Lincoln's participation, this case against Manny made the history books not necessarily for the impact of the reaper patents on the grain industry, but for introducing Lincoln to two key members of his future cabinet, Edwin Stanton and William Seward. Often noted by historians, Lincoln's fee—which turned out to be largely unearned—enabled him to continue his debates with Stephen Douglas, propelling him to become the leading candidate for the Republican Party.

McCormick, as was his style, spared no expense in enforcing his patents. He wanted to teach his competitors a lesson. To lead his fight, he engaged Edward Dickerson, the same patent attorney who handled previous cases for both Colt and Goodyear. To further bolster his chances, McCormick also re-engaged William Seward, who had argued the same patent two years before in New York, and now added former Senator Reverdy Johnson, the man who a year later would represent the slave-owning defendant in the Supreme Court case *Dred Scott v. Sandford*. Although heavily in debt, Manny hired a seasoned patent attorney, George Harding of Philadelphia, and his partner, Edwin Stanton. Manny's team also utilized the talents of other notables, including Abraham Lincoln, Stephen A. Douglas, Peter H. Watson, and sitting Congressman H. Winter Davis. The legal talent involved in a single patent case has yet to find its equal.

In his complaint, McCormick sought an injunction plus damages of $400,000. Initially, the case was to be tried in Springfield, Illinois, by Judge Drummond of the Northern District of Illinois. Watson, lead attorney for

Manny, needed local counsel in Illinois so he retained Abraham Lincoln. Lincoln was well aware of the case and enthusiastically accepted. Watson paid Lincoln a $500 advance and, as was his custom, Lincoln immersed himself in the case, visiting the Manny facility in Rockford and preparing the case for trial.

Much to Lincoln's dismay, the case was transferred to Cincinnati, Ohio. Lincoln wrote to Watson to finalize the details for trial but received no reply. As trial approached, Lincoln continued his preparations, still waiting for word from Watson. None came, mostly because Stanton, Harding, and Watson didn't need, or want, Lincoln's assistance now that the trial had been transferred to Ohio.

With trial looming and no word from Manny's legal team, Lincoln left for Cincinnati. Stanton coldly greeted him, later referring to him as a long-armed baboon who dressed like a creature adorned in a stained suit. Lincoln was shunned until trial, being denied invitations to evening receptions and strategy planning sessions. His carefully prepared brief was never touched by Stanton. Later, Dickerson would report that Lincoln's treatment was nothing short of humiliation and mortification.

Ever the student, Lincoln decided to attend the trial as a spectator on his own dime. He quietly sat in a back seat, taking careful notes, awestruck with Stanton's arguments, which even the judge acknowledged as of "surpassing ability and clearness of demonstration." It was this speech that would later convince Lincoln to replace his original selection for Secretary of War with Stanton.

What many do not realize about the Manny case is that it did not involve McCormick's original patent, which had by this point expired and, according to the court, "belong[ed] to the public." Rather, the case involved two improvement patents that McCormick filed in 1845 and 1853. Of these patents, only three claims were asserted. These were to very specific improvements, such as the position of the raker. The problem with McCormick was that he treated these improvement patents as if they covered every reaper, choosing to ignore the fact that his first patent—which did cover a broad spectrum of reapers—had expired. In so doing, McCormick was among the first to use a corporate patent portfolio to browbeat his competition into submission.

The entirety of Manny's defense was that his reaper didn't infringe. The decision as to infringement was an easy one for the judge. Because patent models were still required, Manny's attorneys used McCormick's model to aptly demonstrate what specific features McCormick claimed as his invention and how these differed from what Manny employed. A simple inspection showed that "the raker on Manny's machines does not require the same elements of combination that were essential in McCormick's invention." In other words, Manny's reaper simply did not infringe McCormick's patents.

McCormick's lawyers raised every tactic imaginable, even conjuring up an argument commonly used today to prove infringement. McCormick's attorneys argued that Manny's design was an equivalent to McCormick's— the counterpart to today's doctrine of equivalents, a doctrine where a patent holder can claim infringement when the infringer avoids the literal language of the patent claim but still sells a device or practices a process that only slightly differs from the claim (referred to as an "equivalent"). As support for this position, McCormick's attorneys pointed to the language in the claims that included the phrase, "or the equivalent therefor."

But in the 1850s, there was no such thing as the doctrine of equivalents, even if the patent claimed to cover equivalents. It was no surprise that the court refused to entertain this argument, saying the claim does not extend "to any improvements which are not substantially the same as those described, and which do not involve the same principle." In other words, Judge McLean refused to find infringement unless Manny's reaper was the same as that claimed in McCormick's patent. When the arguments were finished, Judge McLean held that Manny's reaper did not infringe.

In comparison with today's patent litigation, the case was remarkable for what it did not include. Today, it would be unthinkable for Manny *not* to argue the invalidity of McCormick's patents based on the existence of previous reaper designs. With the ability for today's lawyers to combine any number of prior art references to argue that claims are obvious, any number of invalidity arguments can be raised, escalating the cost of patent litigation by orders of magnitude. But for Manny, there was no discussion about whether McCormick's patent claims were obvious. There was no rhetoric about how any old mechanic could have come up with the reaper by combining known parts from other machines. It was a simple case: McCormick's patent was valid, but Manny's device was different and didn't infringe. From start to finish, the case took less than a year.

But what was the price tag? How much did it cost to hire the most famous legal minds in American history? When Lincoln returned, Watson sent him a $2,000 payment. Honest Abe returned the money, telling Watson that he hadn't earned his fee. But when Watson sent it a second time, Lincoln accepted it, splitting the profit with his law partner, Herndon. Stanton, the silver-tongued lawyer who won his client's case, was reportedly paid $10,000, still less than the $15,000 that Goodyear paid Daniel Webster. Although this amount was thought to be outlandishly expensive and taxing on the court system, when compared to today's patent litigation, the fees were relatively benign. Manny, heavily in debt, was still able to hire the best lawyers in the country without bankrupting himself. With reapers selling at $100 a piece, the cost of the trial was paid off by the sale of 200 reapers.

The adverse judgment didn't deter McCormick. He appealed the case to the U.S. Supreme Court, which took two years to render its decision. The justices agreed with the trial court on all counts, and Manny's victory was confirmed.

Chapter Fourteen

MORSE ENCOUNTERS SALMON P. CHASE

The battle between Goodyear and Day paled in comparison with what Morse faced in Henry O'Reilly. And it took Morse far longer to commercially establish the telegraph than it did for Goodyear to do so with rubber. This may well have been because of the large capital costs required to run wires across America's sprawling countryside. Nobody wanted to take the risk of paying for the infrastructure without assurances that people would send the telegrams. Unable to secure large grants, Morse had to rely on whatever his business partner, F. O. J. Smith, could scrounge up. As Smith was a former member of the U.S. House, Morse continued to hope that Smith's connections could secure some kind of funding, if not from government grants then through private investment.

Morse was beginning to regret his decision to take Representative Smith on board. Each time a new grant was landed, Smith and Morse butted heads over how to spend the money—Morse frugally and Smith extravagantly. In January 1845, when Morse wanted to extend one line to New York, Smith thought they should lay four lines. Beside himself, Morse turned to his old friend Henry Ellsworth for advice. As a compromise, Ellsworth proposed a two-wire line that would cost $50,000, rather than the $100,000 four-wire line proposed by Smith. Ellsworth may have also suggested that Morse cut costs by using Charles Page, the new chief patent examiner, to come up with a crude generator to replace the ever-troublesome batteries. This all upset Smith, who said Ellsworth should stop his meddling and stick to running the patent office.

Morse continued to feel that the best way to operate the business would be to have the government to run his company because of its similarity to the mail business. Morse offered his patents to the government for $100,000 and was politely declined.

After his offer was turned down, Morse knew he needed a change in strategy. In 1845, Morse came to the conclusion that Smith wasn't his man, and that he needed a company, not an ex-politician, to run his business. Morse chose former U.S. Postmaster-General Amos Kendall as his agent. Because Smith still had a twenty-five percent interest in the telegraph patent, Kendall sought Smith's permission to create the Magnetic Telegraph Company. Smith agreed as he would received a quarter of any profits from the new company. Professor Gale, Morse's colleague at New York University who helped Morse with his first experiments, also owned a small interest in the Morse patent, as did Morse's first investor, Alfred Vail. When these men also joined their interests, Kendall now controlled three-fourths of the company. The first project for the new company was to run lines from Baltimore to Philadelphia and New York.

Undaunted by previous failures, Morse in 1846 again sailed to Europe to promote his telegraph. But while he was away, the competition slipped in, determined to succeed where Morse had failed. The copying was bold and brash, with other telegraph companies running their lines side by side with those of the Magnetic Telegraph Company.

The new company knew it had to act to stop the blatant infringement on the Morse patent. But could a single patent wield enough power to shut down this new communication craze, one much bigger than our own Blackberry revolution?

The telegraph was unlike anything the world had ever experienced. Everyone wanted to communicate long distance. It was the original form of texting. By 1850 around twelve thousand miles of lines were being operated by twenty companies in the U.S. alone. Two years later, a cable would connect London to Paris. A mere six years later, the Atlantic would be cabled. Like today, it wrought havoc with social norms. The social eruption was bigger than the Internet. Never before could people communicate in real time over long distances. It was touted as the greatest invention of the generation. But to Henry David Thoreau, all this mindless chatting meant nothing if the two parties had "nothing to communicate." Thoreau felt that what Morse had created was a way to refine human instruments but not human beings. The telegraph was merely an "improved means to an unimproved end."

Still, could Morse's patent shut this down, leaving the Magnetic Telegraph Company as the sole controller of America's lines?

The biggest offender of infringing on Morse's patent rights was Henry O'Reilly, one of Morse's former contractors. The problems with O'Reilly began when Kendall contracted with O'Reilly to put trunk lines to Kentucky and Tennessee, and another line to Cleveland, Detroit, and Chicago. This association proved to be ill founded. Mismanagement and questionable fraudulent behavior by O'Reilly incensed the Morse team, who charged that O'Reilly breached the contract. They then sought an injunction against O'Reilly which was denied for technical reasons—the lawyer named the wrong parties.

O'Reilly took his revenge by seeking to invalidate the Morse patent. In the press, he charged that Morse wasn't the first to invent the telegraph and that Morse should have his patent revoked because he had neglected to sign his application until April 7, 1838, the day after Congress granted Morse his $30,000 to build the first line from Washington to Baltimore. Why this would have invalidated the patent is a mystery. O'Reilly also charged that the Morse patent was invalid for merely claiming a principle—a common practice at the time based on a line of old British decisions.

If none of these arguments prevailed, O'Reilly was confident he could develop a telegraph device that didn't infringe the Morse patent. To that end, O'Reilly found a particularly appealing design-around—a printing telegraph, patented by Royal House, that looked like a piano. Each key of the piano represented a letter that would be printed out when pressed. O'Reilly took a license to the House patent where House granted O'Reilly permission to use a printing telegraph.

In an attempt to block O'Reilly's strategy to design around the Morse patent by using the House printing telegraph, Morse went to the patent office to check out the House patent model. Seeing the model, Morse came to the conclusion that he had invented the piano idea before House, then audaciously filed his own caveat with the patent office to preempt House's patent, claiming that he, Morse, had previously invented the printing idea.

This concept of being first to invent, rather than first to file a patent application, was a part of U.S. patent law that would bog down the system and ultimately protect the rich corporation, rather than the poor inventor it was intended to protect. And it was fraught with fraud, as Whitney had discovered, when later inventors saw patent models, then claimed they were the first to invent. Thirty years after the Morse litigation, Alexander Graham Bell would use the same tactic to secure his own patent on the telephone.

Ignoring Morse's threats, O'Reilly continued constructing more lines, then in late 1847 formed the People's Telegraph Company to build a line to New Orleans. That was enough for Morse to file his first patent suit against him. As expected, O'Reilly argued that Morse was claiming a patent on the general principle of sending electromagnetic signals, rather than a specific telegraphic device. O'Reilly also dredged up some prior art to argue that the Morse patent was invalid.

On September 13, 1848 Judge Monroe rendered his decision in favor of Morse, issuing an injunction. O'Reilly responded by proclaiming that he would appeal his case to the Supreme Court, which he did. But O'Reilly didn't end there. He also claimed that the only reason Morse received his patent was because of a conspiracy in the patent office. He cited Morse's friendship with Ellsworth, the head of the patent office. O'Reilly aptly pointed out that Morse was living with Ellsworth when Ellsworth granted him the patent. O'Reilly called it the "Morse clique."

As is the case today, the best way to get around an injunction is to come up with an non-infringing alternative. Setting aside the House piano key telegraph, O'Reilly came up with the idea of having the telegraph operators write down the messages rather than have a mechanical receiver do the work. The problem was that Morse's patent covered the transmitting side as well as the receiving side of the communication. O'Reilly still infringed the patent every time an operator sent a message. Morse went back to court to enforce his injunction. Fed up with O'Reilly's antics, the judge ordered U.S. Marshals to seize the infringing equipment.

But O'Reilly refused to give up. The telegraph business was just too lucrative to walk away from empty-handed. In 1848, O'Reilly came up with another strategy to avoid Morse's patent. O'Reilly was convinced that he could patent better technologies for sending signals, then use those technologies to compete with Morse. His next foray utilized a telegraph known as the Bain telegraph that recorded the sending message—rather than the receiving message—on tape in the form of dots and dashes. This tape was then used to generate the sending signal.

When Morse got wind of this, Morse went on the offensive and filed his own patent application two months before Bain, saying he had experimented with the idea before. As with the House technology, Morse was able to use the "first to invent" doctrine to derail O'Reilly's plans. Leonard Gale, Morse's former colleague and part owner of the Morse patent, was now a patent examiner and picked up the two cases. Gale held that the Bain patent application "interfered"—claimed the same idea—as Morse and awarded the patent to Morse, rather than Bain. To what extent Ellsworth was involved in protecting Morse is unknown, but his participation was likely. In February1849, the District of Columbia

Circuit Court heard the case and awarded the two men their own separate patents, but limited the claims of each to their specific implementations.

O'Reilly took this as a sign that he could now freely compete. He misunderstood that getting your own patent doesn't give you a right to practice your own patented invention. Even though Bain had a patent on a different implementation, he still infringed Morse's original patent. O'Reilly moved ahead, laying additional lines, and firming up his plans to go to New Orleans.

O'Reilly took his case to the people, proclaiming himself as the "Napoleon of the Telegraph." O'Reilly was dead set on taking over the entire telegraph business in America by stringing lines from coast to coast, including running a line to California. He would call it the "people's highway," a kind of nineteenth-century information superhighway. How he thought this reckless behavior could avoid getting tangled with the Morse patent is anyone's guess. Perhaps O'Reilly thought that if he could run the lines, all of America would stand up for him if Morse and the courts shut him down. It would be a major test for the strength of America's patent system.

Instead of enjoying his newly earned wealth, Morse found he was spending most of his time preparing for trials. Morse was outraged at O'Reilly's claim that the Morse patent should be limited to one specific way of sending letters over a wire. Morse argued against British case law, saying his patent covered any way to send messages over long distances— essentially, that his patent covered a principle. Others, especially those wanting a piece of the booming telegraph market, argued that his patent was only for his specific equipment, used to send and receiving telegraphic signals.

While the fight with O'Reilly continued, the battles with Smith escalated. The problem was that Smith still held an interest in Morse's patent, and he used that leverage to constantly take jabs at Morse. From granting unauthorized lines to stopping communications between Boston and New York (delaying news carried from ships originating from Europe from reaching the New York Associated Press), Smith kept up the fight, though still a part owner of the Morse patent.

The feud reached its climax when a competitor, Hugh Downing, built a line from New York to Boston to compete with Smith's line. Smith chose to put Morse's patent in play, suing Downing in Boston. The case was assigned to Judge Woodbury in Boston, the same judge who heard Colt's case. Downing was using House's piano equipment, and like O'Reilly was prepared to argue that this technology didn't infringe. Morse, having won a previous patent infringement suit, thought the case would be an easy one for the judge.

But that didn't happen. In October 1850, Woodbury admitted that he didn't really understand the technology, then held that Morse's patent was for his machinery and his code, not for the general principle of electrographically sending messages. Thus, the House transmitter did not infringe Morse's patent. This stunning defeat must have surprised Morse, especially after Woodbury had been so pro-patent with Colt, essentially assuming any firearm that shot in rapid succession was an infringement of Colt's patent.

For Morse, the way the case was argued was of far greater importance than the judge misunderstanding scientific principles. The loss of the suit, according to Morse, happened primarily because Smith and his attorney had argued the case poorly. Smith, in an act of petty revenge, chose not to have his attorney challenge the defense witness's testimony simply because Morse had refused to testify at trial. Not crossing a defense witness is like telling the jury you have no confidence in your case, but Smith refused to cross examine the witness unless Morse would cooperate and rebut the defense with his own testimony. Jackson's testimony (the other person on the *Sully* who claimed he came up with the idea of the telegraph before Morse) stated that Morse hadn't invented anything because at the time Morse was clueless as to the principle of electromagnetism, and that he, Jackson, was the true inventor that day on the *Sully*. The testimony was so insulting to Morse that he refused to even combat it. These bizarre positions—Smith refusing to rebut the defense and Morse refusing to testify—convinced Woodbury that Morse really didn't invent the principle and that Morse's patent claims should be limited to the specific equipment used to transmit and receive the signals.

After the case, the tenuous relationship between Morse and Smith reached an end as Smith was accused of being incompetent for refusing to have his attorney cross examine the witness. Perhaps to seek revenge, Smith severed his ties with Morse and joined O'Reilly. So Morse repaid the deed and sued Smith in New York for patent infringement. As part of the New York case, Morse claimed that when trying the Downing case, Smith instructed his lawyer to argue for the other side just to ensure that Morse would lose—and Smith joining forces with O'Reilly after the case proved the point. All along, Morse claimed, Smith had a plan to defeat Morse's patent so he could compete without interference from Morse's patent. In essence, Morse was arguing that Smith was attempting to defraud Morse out of his patent. Rightly, the judge refused to grant the injunction against Smith.

With O'Reilly still in the market and gaining strength every day, Morse became disillusioned with the American legal system. He was spending all his time and money on lawsuits and yet had accomplished little in terms of stopping the competition.

To some extent, Morse himself was to blame. He could have easily chosen to take the witness stand to bolster his own cause. In part, then, Morse's situation was not due to the patent system, but Morse's inability to get along with Smith.

The good news for Morse was that he had one last chance to redeem himself. Morse had taken his case against Smith and O'Reilly on appeal to the Supreme Court.

The Great Telegraph case reached the Supreme Court in December 1852. Justice Roger B. Taney presided, the same judge who later decided the *Dred Scott v. John Sanford* case, where slaves were not considered to be "persons" under the Constitution. Suspicions of nepotism were raised as Taney had just received an honorary LL.D. from New York University, the same institution where Morse had constructed the first working telegraph.

O'Reilly strengthened his position by engaging Salmon P. Chase of Ohio—another member of Lincoln's future cabinet—to argue his case. Chase raised all the same legal issues that were made in the lower court case, including that aboard the *Sully*, Jackson, not Morse, was the first to come up with the "principle" of electromagnetic communication. And if not Jackson, then a dozen others. On top of all that, Chase argued that O'Reilly's telegraph was so different that it didn't infringe in the first place.

The decision for the Court to make came down to two issues. Was Morse the first and original inventor? And, was O'Reilly's telegraph "substantially different" so as to not infringe?

Chase argued that telegraph machines had already been invented by others, primarily Charles Wheatstone. Morse may have made the first practical "marking" telegraph, but that was it. Morse had done nothing more than cobble together known ideas, such as transmitting electric currents over wires, using batteries as a power source, and using electromagnets to move letters.

Then Chase argued that the O'Reilly telegraph was an improvement, not an infringement. In essence, Chase was arguing that improving on Morse's patented design gave O'Reilly immunity from Morse's patent. This is similar to arguing that being awarded a patent on an invention gives you the legal right to use that invention. Even the brightest minds can mess up. Legally, Chase was wrong. Even if O'Reilly had an improved telegraph, it still operated using principles claimed in the Morse patent.

Chase grumbled about Morse's abuse of the patent system, extending the breadth of his claims through the "reissue" practice—the same practice Colt had been guilty of in extending the life of his revolver patent. Although not successful, Chase highlighted a problem with the patent laws, one that we are still living with today: the uncertainty of patent rights when the patent holder can keep going back to the patent office to get additional patent claims. If Colt and Morse abused the system, it would be nothing compared to what the future held in store, when George Selden held the entire automobile industry hostage.

It took until February 1854 for the Court to render its decision. In anticipation of the announcement, *New York Times* said in its commentary that it was only fair to hold in favor of Morse, not just because he had a patent, but because Morse had almost singlehandedly developed the market.

The Court agreed and found the patent valid, admitting that while Morse may have borrowed ideas from others, that was within his rights. After all, every invention is a combination of old ideas.

Chase was unfortunate. He was a hundred years ahead of his time. Those same arguments—that Morse had merely cobbled together old ideas—if presented today would have easily invalidated Morse's patent. Unlike the Court of the 1850s, today's Supreme Court is hostile toward patents, essentially agreeing with Chase's line of reasoning in its most recent decision on the issue of obviousness, now referred to in the patent profession as the *KSR* decision.

The justices also found infringement and gave a perpetual injunction. The Court held that the O'Reilly telegraph essentially did the same thing as claimed in Morse's patent. In other words, it didn't do something differently or leave something out. Chase's "improvement" argument— that O'Reilly had an improved design—wasn't a very good argument. An enhanced design can still infringe somebody else's patent.

The only good news for O'Reilly was that the Court struck down the patent's eighth claim—as to the principle of using current to mark or print characters—as being too broad. But this was all a moot point. An infringer need only infringe one claim to infringe the entire patent. The injunction still held. The Court's decision was so sweeping that the House printing telegraph and Bain's electrochemical version were both considered to be infringements. After five years, Morse essentially put O'Reilly out of business. O'Reilly's company filed for bankruptcy, leaving O'Reilly to take a clerking job in New York. To add to the insult, Congress granted Morse's request to extend his patent an additional seven years.

Chapter Fifteen

THE AFTERMATH

During the 1850s, America's economy was back on track. Its factories were singing the praises of its great inventors, household names like Morse, Goodyear, and McCormick. These men—not the politicians—had taken America far beyond an agrarian society.

The three great Supreme Court cases concluded, their victors were ready to go down in the annals of history. McCormick, Morse, and Goodyear were just as vital to America as were Madison and Franklin. To memorialize their sacrifices, a tribute was painted on the Capital's canopy—The Apotheosis of Washington—completed in 1866. It represented the passing of the torch, from Washington to the great innovators who had transformed America. The artist, Constantino Brumidi, made sure nobody would overlook their contribution as he painted his great canopy fresco.

In Brumidi's masterpiece, Washington makes his heavenly ascent to be crowned as one of the gods, leaving America, his creation, behind. But he doesn't leave his people alone or helpless. He has already transferred his mantle, not to politicians, but to the great innovators. And to make sure they do not fail, he leaves behind a few gods to oversee the work of inventing.

The fresco shows Washington at the apex, rising to heaven in glory, accompanied by thirteen maidens, symbolizing the original states. Beneath him, six groups of figures line the perimeter of the canopy, representing the six technology areas needed to make America great. Each is presided over by a carefully chosen god, making sure America's inventive talents will propel it to its Manifest Destiny.

In "Agriculture," Brumidi highlights McCormick's reaper, laden with sheaves of wheat, being ridden by Ceres, the goddess of agriculture. "War" depicts a figure representing Armored Freedom banishing a sword, perched over a canon. Missing from these instruments of war is Colt's revolver, possibly because of Colt's tarnished reputation. Minerva, the goddess of wisdom, presides over "Science," glorying over an electric generator and batteries for storing power, while Franklin, Morse, and Fulton watch with an eye of satisfaction. It was this generator that created the power to drive Morse's telegraph. In "Marine," Neptune, god of the sea, watches as a naked Venus, goddess of love that was born out of the sea, lays Morse's transatlantic cable. And representing Goodyear's rubber-processing invention is "Mechanics," where Vulcan, god of the forge, stands atop a canon, adjacent an anvil, with a steam ship behind. It is as if Vulcan is lending his name to Goodyear, putting his stamp of approval on the revolutionary vulcanization process.

The fresco was nothing short of complete vindication for Morse. Although the Rotunda was devoid of any of his paintings, the great ceiling masterpiece showed not only the laying of his transatlantic cable, but also included his own a portrait as he inspects the generator used to create his electromagnetic transmissions.

The Supreme Court's decision brought a huge windfall to Morse. Virtually all U.S. telegraph companies capitulated and paid royalties on Morse's patent. And in 1857, Congress extended his patent another seven years. By this time, Morse had already made a handsome profit and was clearly not entitled to the extension under existing law. To justify his case, Morse used some creative accounting and calculated a low-balled profit of only $200,000 over the life of his patent. Usually, even this amount was enough to deny the extension, but Congress apparently thought Morse was entitled to a little extra.

It was enough to let Morse live out his days in leisure. Now married to his second cousin (who was twenty-six years younger), the couple had seven children and lived in their mansion, Locust Grove, that overlooked the Hudson River.

Meanwhile, the telegraph business began to consolidate. In 1856, the New York and Mississippi Printing Telegraph Company merged with several other companies to form the Western Union Telegraph company. Then, a decade later, Western Union merged with Morse's telegraph company, the American Telegraph Company, and proceeded to dominate the U.S. market until the telegraph became obsolete.

In a fitting tribute, Morse sent what would be his farewell telegraph on June 10, 1871. It was the first telegraph to traverse the entire world. A year later, Morse died at the age of eighty.

McCormick also turned out to become one of America's heroes, though his future was far from certain following his loss to Manny. The reaper market was now wide open. The Supreme Court had spoken—McCormick's patent didn't cover every reaper.

What may have surprised McCormick is that even after losing his case against Manny, McCormick's infringement problems began to wane. Even the hated Hussey was no longer a threat, even though Congress had granted Hussey a patent extension. This was in large part because McCormick managed to beat most of his competition on the basis of performance, not his patents.

Hussey's business had been in decline ever since his final duel with McCormick at the 1851 industrial exhibition at the Crystal Palace in London—the famed London World's Fair. As reported in the papers, large crowds gathered on a wet July day to watch as McCormick handily beat the Hussey in a field of green wheat. It was impossible for Hussey to stop the news from spreading. In reporting McCormick's win, the

London Times wrote that "the reaping machine from the United States is the most valuable contribution from abroad, to the stock of our previous knowledge, that we have yet discovered."

Tired from years of struggling to compete with the likes of McCormick, in 1858 Hussey sold his reaper business and patent for $200,00 and went to work on a steam plough. Two years later, while traveling from Boston to Portland, Maine, the train he was riding ran short of water. When the train stopped to take on additional passengers, a child asked for some water and Hussey volunteered to find her some. As Hussey re-boarded the car the train unexpectedly started, and Hussey was accidentally thrown beneath the wheels, ending his dream for fame.

Despite his loss to Manny, McCormick was smart enough to not stop inventing. If he could improve the reaper, so could others. McCormick knew all too well that patents last only so long. If he wanted to stay one step ahead, he needed more of them. McCormick's genius was to sense the pulse of innovation and realize that America held significantly more inventive talent than just his own. McCormick's greatest pool of potential inventors were men just like himself—everyday farmers.

While factories were in the business of blindly mass producing reapers, farmers—the ones who used the reapers—were making thousands of improvements. They preferred new designs from their fellow farmers rather than from a corporate salesman. This put pressure on the manufacturers to incorporate the field-invented designs into their own reapers.

Although most of the improvements were made by poor farmers and small-town mechanics, that didn't stop them from spending money to file patent applications on their designs. That meant that manufacturers like McCormick couldn't just willy-nilly copy the most popular reapers of the season. Instead, McCormick sent out "patent scouts" to scour the country for new designs. For those inventions worthy of being commercially implemented, these patent scouts bought up the patents for McCormick, who used the winter months to incorporate the designs into his own machines.

McCormick also put pressure on his mechanics to come up with improvements. When they did, McCormick filed its own patent applications. But unlike the rural farmer who was entitled to royalties for patents used by others, employees were required to assign their patent rights to their employers. And thus started the decline of innovation, when the patent system became the slave to the corporation. Ironically, with this action it was the inventors that now became mechanized, rather than the machinery.

Although McCormick had the resources to fully utilize the patent system for his own benefit (and he tried to do so on many occasions), the patent system had not yet evolved to include all the procedural advantages that would later be used to smash smaller competitors. Instead, reaper technology continued to evolve, moving from the mower to the self-rake reaper, then to the harvester, and finally to the harvester-binder. And McCormick hardly had a monopoly. In the 1860s there were about fifty independent reaper manufacturers, and the country had more than 50,000 operating reapers. This replaced the work of 350,000 men and saved $4,000,000 in wages. Amazingly, these machines harvested 50,000,000 bushels of grain. By the time the Civil War ended, the number of machines had reached 90,000. The field evolved so rapidly that by the twentieth century, the horse-drawn reaper was outmoded by the steam-powered combine that simultaneously cut and threshed the grain. Yet it too borrowed from McCormick's original design.

For Goodyear, the end of his battle with Day didn't make life any easier. Ironically, Day became wealthy while Goodyear remained in dire poverty. Day may have been shut out of the shirred goods market, but his factories along the Eastern seaboard kept churning out other rubber products, netting him more than a million dollars and enabling him to set himself up in high style in Manhattan. When Goodyear sought his patent extension in 1858, Day was there to challenge him. To obtain the extension, Goodyear had to prove that he had not realized a fair return on his invention. Although he had grossed $162,894 from licensees over the life of his patent, his expenses gave him a net profit of only $33,000. This was remarkable in that his licensed shoe manufacturers were selling over five million rubber shoes a year.

Day, of course, went on record saying that there was nothing original about Goodyear's patent, and it was only due to mismanagement that he couldn't turn a decent profit. He also didn't believe that Goodyear's numbers could possibly be accurate.

Even if Goodyear had made enormous profits, he might well have gotten his extension anyway, mostly because the patent office hated Day. Ellsworth had caught Day trying to claim Goodyear's vulcanization process more than a decade before. The hatred between Day and the patent office escalated in 1849 when Goodyear applied to have his patent reissued. In the five years after his original patent had issued, Goodyear had learned that vulcanization was produced by heating rubber and sulphur, without the need for white lead. Goodyear therefore sought to broaden his patent claims to cover virtually any vulcanization process.

The patent examiner handling Goodyear's case was none other than Dr. Gale, the same professor who helped Morse develop the telegraph at New York University and who was accused of showing favoritism to Morse when granting Morse's patent ahead of Bain's application.

For Day, this was déjà vu. He couldn't see the justice in letting patent holders continuously broaden their patents, then get patent extensions. Day had a legitimate point. The problem was how he went out about voicing his opinions. Day wrote a public letter claiming that men such as Gale were not fit to work in the patent office or to protect the sacred rights of inventors. If Day hadn't been such a scoundrel, he might have made headway with his arguments. As it was, the patent office turned against him.

Goodyear was awarded his seven-year extension. Joseph Holt, the new head of the patent office, rendered his decision by recounting Goodyear's perseverance in the direst of circumstances, reliving his trips to debtor's prison. Holt overlooked Goodyear's poor bookkeeping, assuming that with Goodyear's penury, he couldn't possibly have profited from his patent. In the end, Holt found that he just couldn't deny the extension. "I should be false to the generous spirit of the patent laws, and forgetful of the exalted ends which it must ever be the crowning glory of those laws to accomplish."

The patent term extension was perhaps Goodyear's most fulfilling moment. Deeply in debt, he desperately needed more royalties. Following his initial victory over Day and the immense success at the 1851 London Exhibition, Goodyear decided to go all out in Paris for the 1854 World's Fair. Goodyear went extravagant, spending $50,000 to create every kind of rubber article imaginable, from sofas to eyeglasses. His display was so grand that it even captured the attention of Napoleon III, who invited Goodyear to join him on his carriage rides through Paris. Adding to the money drain, Goodyear purchased a home on the Champs-Elysees.

The end result was that Goodyear ran short on cash, landing him in the Clichy Prison in Paris. In a single year, he'd ridden with an emperor and been thrown into jail.

Before returning home, Goodyear put himself through one more lawsuit, this time in England, trying to convince a court that his vulcanization process was stolen by Thomas Hancock, who reverse-engineered some of Goodyear's rubber samples and then beat Goodyear to the British patent office to patent the idea. Goodyear won the sympathy of the judge but not his verdict. England had a "first to file" rule, and Hancock had beat Goodyear to the patent office.

Though Goodyear returned home in 1858 to the good news of his patent extension, any sense of relief would be short lived. His health was broken, and he resorted to hobbling around on a cane. He died in 1860, $200,000 in debt. Not until years later would his son, Charles Jr., see the financial rewards of Goodyear's patent. But even then, Charles Jr. was forced to face Day one more time when he applied for a second seven-year extension to his father's patent. Day testified in opposition, dredging up all his former arguments, and harshly cross examining Charles Jr.

But others also opposed the extension, including the railroads and other industries that relied on rubber parts. This time, the patent office denied the request. A twenty-one year monopoly was enough.

———

Two cases never made it to the Supreme Court—those involving Colt and Singer. Not surprisingly, neither man made Brumidi's painting. Not including Colt's revolver in the "War" scene or Singer's sewing machine in "Mechanics" could only have been an intentional shun. The brash personalities of these two men all but assured their absence from the Capitol's canopy. Still, Singer and Colt both became millionaires as the sewing machine and revolver industries boomed.

Colt's brashness continually threw him into the public light. He was an admired figure among the rough-shod men who bought his revolvers, the forty-niners, the Texas gunmen, and all kinds of military officers. But for mainstream America, he was too politically incorrect—a man with a secret bastard son and an insane brother, a radical on the slavery issue (saying that it was an inefficient use of labor). His image was further tarnished in 1854 when Colt tried to get a second patent term extension, which had already been extended in 1849. All told, Colt was trying to get 28 years of patent life. Allegations were raised that he tried to bribe members of Congress with a $15,000 slush fund managed by his patent attorney, Edward Dickerson. Colt was eventually exonerated, but his patent wasn't extended. And Colt never changed his ways. A few years later, Colt again offered $50,000 to a lobbyist in an effort to secure a reissue of an important patent relating to a rack and gear-loading lever that he wanted to include in his revolvers.

Colt never shunned scandal. If anything, he sought it. The publicity it created only helped him sell more revolvers. A Congressional investigation could only work to his benefit.

And the orders came pouring in. He quickly outgrew his small armory in New Haven, Connecticut, where he manufactured the firearms that were rushed to the front lines of the Mexican war. After another temporary move, Colt built his dream factory, a gigantic plant in Hartford that housed operations for the Patent Fire Arms Manufacturing Company. Some estimated that Colt spent in excess of $400,00 on the facility. Capped with a blue dome, it became a famed landmark, changing the skyline of Hartford.

The now-famous Colt revolver patent finally expired on October 24, 1856, and Colt faced stiff competition from the likes of Massachusetts Arms, Remington, and Eli Whitney, Jr. But Colt managed to stay ahead, mostly because of his skilled workers and high-tech factory. In his South Meadows armory, Colt demanded hard work and perfection, driving his machinists ten hours a day, enough to churn out 150 finished revolvers a day by 1857. If his men weren't clocked in by seven, they were locked out until noon with no pay. As an employer, Colt was stern, yet fair, even building his workers homes in a nearby compound that also included a library and community center.

The same year that Colt's patent expired, Colt decided to set his wild, bachelor life aside and marry Elizabeth Hart Jarvis. Like most things in which Colt had his hand, the wedding was completely overboard. Colt hired a steam ship to take him to the ceremony. Following the wedding, the newlyweds left for a six-month honeymoon in Europe, being the special guest of Russia's Czar Alexander II at his coronation. Of course, Colt always kept business at the forefront. He left the coronation having landed a contract to set up an armory in Russia.

If Colt knew how to do one thing, it was spend money. His factory was taking in more than a quarter million dollars a year, charging $24 for every revolver. Ironically, Colt had lost this same amount after his patent had been granted in 1845. Now, opportunities for making money were virtually endless, especially when hostilities began brewing between the North and South. Colt accelerated production and began running his factory night and day. Under the Militia Act, Colt was required to provide arms to any state who requested an order as long as the U.S.

Army's Ordinance Department approved them. And so he did, freely selling to the Southern states, even though Colt was a great proponent of the North and the abolitionist movement.

Colt's death in January 1862 prevented him from cashing in on Civil War sales, though his wife certainly did, becoming the richest widow in New England. With the additional sales during the war, a year later Colt's Armory was selling enough firearms to net the company millions of dollars.

———————

The scandals surrounding Colt paled in comparison to those Singer brought onto himself. With the Great Sewing Machine Combination, Singer no longer had to worry about being harassed by patent infringement cases, although he did have to pay into the Combination longer when Congress granted Howe a seven-year patent term extension. Singer devoted less and less time to his business, and more and more time tending to his complicated personal life.

During his lifetime Singer was reported to have had six common law or legal marriages, some at the same time. In 1860, his two most recent families both lived in New York City. For these, Singer took on different names and pretended he knew nothing about Isaac Singer or the sewing machine. Singer's luck ran out one day when he decided to take a carriage ride with one wife, Mary McGonigal, down Fifth Avenue and just happened to pass Mary Sponsler, Singer's other New York wife. Upon seeing Singer with another family, Sponsler threw a tantrum in the middle of Fifth Avenue, driving Singer's uncontrollable temper into a rage. Singer waited for her at their home, intent on beating her into submission—or at least making sure she'd never breathe a word about what she'd seen. When Sponsler's daughter came to her mother's defense, Singer beat her instead, leaving her unconscious.

To avoid prosecution, Singer fled to Europe, where he wooed another bride-to-be, Isabella, who eventually bore him six more children. In

1863, Singer returned with Isabella to New York, where they married—only after Singer's bribes and threats of violence convinced Catherine to grant Singer a divorce. While he was there, Singer turned over the reins of his sewing machine operations to Clark, who formed the Singer Manufacturing Company, virtually ending Singer's association with the company that bore his name.

Singer quickly discovered that New York society wanted no part of him or his new bride. He returned to Paris, then to a seaside resort near Devon until he passed on July 23, 1875.

Clark, following Singer's advice, dominated the sewing machine market by targeting home sales. He did this by reducing the weight of the sewing machine from over a hundred pounds to a weight that could be easily lifted by a woman. Clark also latched onto McCormick's financing strategy and offered a machine for $5 down and $3 per month for 16 months. When the patent pool expired in 1877, Singer had captured half the U.S. market, selling over 250,000 machines a year.

Chapter Sixteen

THE OLD CURIOSITY SHOP IS MOTHBALLED

In 1836, the dream for the new patent office was that it not only be fireproof, but that it serve as a Museum of the Arts. What eventually transpired went beyond Dr. Thornton's dreams when he proposed replacing Blodgett's hotel with a new patent office to showcase all of its models. The patent office turned into not only a place to examine patent applications but into the National Gallery. In addition to the nation's inventions, it housed many of America's treasures. Along the way it served as a hospital for Union soldiers, then as the dance floor for the Inaugural Ball at Lincoln's second inauguration.

Now housing the National Portrait Gallery, the building has three stories, with four halls that surround a marble-tiled inner courtyard, affording visitors a floor-to-ceiling view. Each end terminates in a large portico, one being a replica of the Pantheon in Rome. In the 1870s, the Model Room occupied the entire third floor.

The models were displayed in cases, stacked two high, to provide enough space to show off around 200,00 models. Each case was made of white pine, with glass walls and sides so that the models could be viewed from any angle. For extra-curious onlookers, the cases could be opened, under supervision, so that the models could he handled.

The museum held so many models that a guide book was published to guide visitors through the maze of exhibits. Upon entering the National Gallery, patrons were treated to Ben Franklin's printing press, famous U.S. treaties, and the personal effects of George Washington, including tents and blankets from his Revolutionary War camps. Also not to be missed was Andrew Jackson's coat worn during the Battle of New Orleans, the hat worn by Abraham Lincoln during his assassination, and the original Declaration of Independence.

But most came to see the models, now part of America's heritage, treasures like Abraham Lincoln's riverboat model. Most were ordinary ideas: musical instruments, medical devices, all kids of eyeglasses and corsets, industrial machinery, engines, sewing machines, and railcars. The savvy tourist quickly learned that a visit to the patent office museum required the services of an experienced guide to lead them through the folding hoopskirt models, railroad cars and other contraptions. Over the decades, millions perused the cases of models.

Even with the immense popularity of the models, hints that their days were numbered began as early as the 1850s when then-commissioner of the patent office, Commissioner Ewbank, began promoting the publishing of patent specifications and drawings, arguing that written documents were easier to disseminate beyond the walls of the patent office, and that they eliminated the need to build an invention that had already been invented. The cry was to modernize the patent office, and paper drawings were the way to start. This was coupled with the reality that with the rapidly accelerated pace of patent filings, the patent office would soon be unable to hold them all. The Secretary of the Interior, which oversaw the patent office, in 1852 urged Congress to approve publication of patent specifications and drawings "in much the larger number of cases, [so that] the necessity for preserving and displaying the models would be obviated." A year later, drawings and abstracts of issued patents were published in the Annual Report.

The demise of the patent model came one step closer in 1870, when a new Patent Act granted the Commissioner the responsibility for requiring models, in essence making them optional based on the discretion of the Commissioner. The next year, when the Commissioner was authorized to print full copies of newly issued patents, the models were nearly obsolete.

Then disaster struck.

When the previous patent office burned in 1836, the politicians vowed to build a magnificent building that could never burn. The problem was that they didn't follow through with their original plans. While the original wings were built with iron and brick, the west or Ninth Street wing used wood trusses and pine sheathing covered by a thin sheet of copper. To make matters worse, the 12,000 rejected models that were stored in the west wing were made primarily of wood—a perfect collection of kindling.

On Monday, September 24, 1877, fire broke out in the west wing, billowing clouds of smoke through the skylights. The rejected models were ablaze, with the fire quickly spreading. Firefighters were called, but those first on the scene were unable to deliver water to the 80-foot roof because the water pressure permitted only a 60-foot stream of water from the hydrants. Other fire teams from Alexandria and Baltimore were summoned to assist.

Risking their lives, employees scurried about, saving what documents they could. Grabbing anything of value, workers hauled their bundles through the hallways and to the lower floors. Perhaps the only thing that saved the building was a brisk south breeze that kept the fire from spreading to the south hall. This allowed the firemen to gain the upper hand, and eventually the fire was doused without destroying the bulk of the building.

The volunteers managed to save 777 folios of drawings containing 211,243 original drawings. That was the good news. Nearly all of the 114,000 models in the west and north halls were destroyed or significantly damaged. Around 40,000 patents were destroyed or damaged.

Then came the tedious task of attempting to restore the models. Congress appropriated $76,000 for the task. This required examiners to sort through the rubble, piling the salvageable models into specific classes. Using patent drawings, the examiners identified the inventions, then tagged them with labels indicating the inventor, patent number, and name of the invention. Cleaners took off the soot and rust using sulfuric acid, then neutralized the acid in a lime water bath. Often, dismantling was required, and machinists were employed to produce spare parts or repair what they could, often bending back metal parts that had been disfigured from the heat. Using their best efforts, the restoration crew managed to save only 27,000 of the models.

The devastating fire signaled the end for the models. Commissioner General Ellis Spear wrote as follows with regard to the models:

> They were useful to the public for the purpose of examination by persons desiring to be advised as to the novelty of inventions made by themselves or their clients. Duplicates were sometimes ordered to be used as evidence in court, and they were frequently referred to in applications for reissue. [But] they form no part of the patent when issued. The law makes it essential to the validity of a patent that the specification and drawing thereof shall disclose fully the invention to those skilled in the art to which it pertains. [Therefore,] examinations can be made without models ... and with greater certainty of fullness and accuracy from the drawings....
>
> under certain circumstances, it will be better to dispense with the models in applications for patents.... It will be necessary only that provision be made for requiring models in cases where the capability of the machine to operate is called into question, or where the Examiner is in doubt as to the sufficiency of the drawings, or where models may be necessary for ready illustration on appeal, or interference cases.... Now

that the models which remain cover half the ground, it seems an opportune time to change the system.

That was the official statement. Off the record, he declared: "We will have no more models. Drawings will do—and the smaller the better."

Thus, in 1880 the patent office discontinued the model requirement except for perpetual motion contraptions and flying machines, which were thought to be an impossibility. Not surprisingly, the model requirement for flying machines was discontinued after the Wright Brothers' airplane invention.

Many reasons were put forward to justify the decision to do away with the models. Models cluttered offices, hallways and even the display rooms. Examiners used them for paperweights, or to scare away the cats at night. While space was the obvious issue, there were others. For one, the patent office was woefully behind in its duties and Congress refused to appropriate more funds. Money used to house the models could be used elsewhere.

The order to dismantle the model collection was equally the result of pressure from large businesses who felt that filing patent models was too cumbersome and onerous. Modern efficiency was what the patent office needed.

So down came the order to get rid of the thousands of models that had been viewed by millions.

The problem faced by the patent office was what to do with them. Since the inception of the model requirement, an estimated 246,094 patents had issued, and 200,000 of those were accompanied by models. Because the Old Curiosity Shop was one of the most popular tourist attractions in the nation's capital, patent office officials knew they couldn't just toss them in the dumpster. Or could they?

At first, patent office workers in 1893 bundled up 155,00 of the models into 2,700 crates, where over the decades they bounced around, from barns to basements, until being dumped in an old livery stable. There they sat until 1925, when the government, having already paid more than $200,00 to store them, refused to allocate any more funds. At that point, Congress allocated a final $10,000 to form a committee of three "do away" with them by any means. Fortunately, the Smithsonian took 2,500 of them—at least the ones they believed to have the most historical value. It was no surprise that the Smithsonian wanted Elias Howe's sewing machine, as well as sewing machines invented by Allen B. Wilson in 1850 and 1852, and James Gibbs in 1857, mostly because they told the story of Howe and the patent wars that ensued between dozens of sewing machine companies.

Other prize models grabbed by the Smithsonian included the two gasoline engines invented by G.B. Brayton in 1872 and N.A. Otto in 1877, which eventually found their way into the automobile and the airplane. Both were also used as evidence in the bitter patent war between George B. Selden and Henry Ford. The courts eventually held that Selden's patent covered an automobile with a Brayton type engine, whereas Ford's cars used an Otto engine.

A large portion of the models were sold at auction to Sir Henry Wellcome, who had plans to create a patent model museum. That all ended with the crash of 1929. Since then, they have passed hands dozens of times, with some ending up in museums, others on eBay.

The committee did take efforts to return some models to the inventors or their families and to award them to cities and museums. The rest that weren't worthy of being purchased at auction were simply tossed, and with them went an important part of America's history.

In 1890, only 535 models were submitted while 26,000 patents issued. Many patent office veterans regretted the decision to no longer require models. Then-Commissioner Mitchell said that it was a public calamity that the model requirement had been suspended, merely for lack of space to store and exhibit them.

Eliminating models meant inventors were no longer required to build their inventions. This change led to the proliferation of modern-day "paper patents"—ideas never built but for which patents have been granted. Not only do these clog the patent office with dubious applications, but when they issue they are responsible for much of today's spurious patent litigation.

A flood of problems began almost simultaneously with the discarding of the rule. If the model requirement had been kept in place for only a few more years, the history of the automobile and the telephone would be vastly different.

Chapter Seventeen

THE TELEPHONE AND THE AUTOMOBILE

The Centennial Exposition of 1876 was held in Philadelphia, a celebration of America's one hundredth birthday. America wanted it to rival the Great Exhibition of 1851 held in London, where Americans like Goodyear, McCormick, and Colt showed off their wares. The patent office joined in the festivities by sending 5,000 of its more famous models.

Alexander Graham Bell, still unknown, presided over a demonstration of his new telephone. The exhibit went mostly unnoticed, still an unproven technology. That same year, Bell's patent had issued, but it too had failed to generate any interest. Enthusiastic and optimistic, Bell may have been the only person with an inkling that his telephone patent would turn out to be the most monetarily successful patent of all time, a patent that gave birth to one of the largest companies in the world's history: AT&T. It also marked the high water mark for the U.S. patent system in terms of encouraging innovation. In the century since, the nation's economy has relied on the success of large corporations. As these corporations learned how to exploit patents, cries sounded for patent reform. The main complaint was that patents were too easy to obtain and too powerful in the hands of controlling companies. And so reform did come, initially through judicial decisions, but eventually codified into law. But in their efforts to reform, the legislators and judiciary have made it nearly impossible for the solo inventor to protect his inventions.

The issues facing the patent system began to become more apparent when on February 14, 1876, Bell coincidently—or not so coincidently—filed his telephone patent application on the same day as Elisha Gray filed his. Still in his twenties, Bell was working on a way to send multiple telegraph signals down the same wire. Anyone strolling the streets of New York City understood the magnitude of the problem, as unsightly telegraph lines wove a ceiling of steel above the bustling city. When Bell discovered a way to send multiple telegraph signals down a single line, much of this unsightly mess could be eliminated. Solution in hand, Bell had his patent attorney rush to the patent office and file his patent application, allegedly just hours before Elisha Gray submitted his own patent application for an "instrument for transmitting and receiving vocal sounds telegraphically."

Bell's application posed a problem. His application had a paragraph, hand-written in the margins, that described the telephone in almost identical language to Gray's application. Some historians now believe that, based on Gray's own testimony, Gray filed his application a few hours before Bell. Somehow, they say, Bell managed to see Gray's application right after it was filed and noticed Gray's idea about transmitting voice signals electronically. The theory contends that Bell then copied the relevant paragraph and handwrote the same material in the margin of his own application, in essence stealing the idea from Gray.

Without requiring Bell to submit a patent model at the time of his filing to prove that he had actually built the first telephone, or determining whether Gray was really the first to invent the idea, the patent office issued Bell's patent on March 7, 1876. Why Gray's application was not considered by the patent office has been debated ever since. The most common explanation is that since Bell allegedly filed his application first, the patent office used its discretion and awarded Bell the first telephone patent. However, this ignores the contention that Gray more than likely filed his application several hours before Bell. Even then, when two applications were filed in close proximity and claimed the same idea, the patent office was required to investigate to see who was the first to invent, not the first to file his application. The patent office failed to do this.

Bell's patent also issued in less than a month—and three days before Bell had even built his first telephone and made his famous summons to his underling: "Watson, come here. I want you." Why the patent office issued Bell's patent so quickly, and without the inclusion of a model, has never been settled.

Even with all the intrigue, Bell's hastily granted patent withstood hundreds of attacks. Congress in 1885 commissioned a special investigation. The resulting report discovered many of the patent office irregularities and concluded that there was overwhelming evidence of foul play, including the patent office decision to give Bell's application priority over Gray's application. Yet nothing came out of it, possibly because several members of Congress stood to benefit financially if the Bell patent remained in force. Western Union also challenged the validity of Bell's patent. The parties ultimately settled with a technology transfer arrangement, and Bell's stock instantly doubled. Bell soon became a telephone monopoly providing telephone services to millions.

What would happen if Bell filed his patent application today?

First, because Bell and Gray filed on the same idea, Bell's patent application would be hung up in a patent office proceeding for about ten years while he fought with Elisha Gray about who first invented the idea. But proving that he was the first to invent the telephone would be the least of Bell's worries. His next hurdle would likely be insurmountable: The patent office's position would be that Bell's idea was obvious, that his breakthrough wasn't inventive enough to rise to the level of being awarded a patent. Why? Because Bell didn't make an inventive leap that was far beyond what was already known in the scientific literature. As one example, in 1861 a German scientist, Philipp Reis, invented a telephone that faintly transmitted the human voice. Reis demonstrated his idea by talking into a speaking tube so that his voice vibrated a metal diaphragm. This modulated electric current from a battery and varied the resistance of the circuit. Although Reis did not vary the resistance by placing a wire into liquid as Bell proposed, today's patent office wouldn't see a patentable difference between the two concepts; both varied the resistance of the circuit. Today, Bell would find himself arguing with the patent office for years, and most likely would be denied his patent.

And what about the telephone during all this legal haggling? After all, the whole goal of the patent system is to provide an incentive to promote new technologies. Most likely, the telephone would have languished in Bell's laboratory, lacking any private investment to commercialize the concept. Then, as now, investors would have been unwilling to put millions of dollars into a technology that they may not end up owning, or in which a copyist could weasel in and undercut the price without expending the same amount of research and development money. Not knowing what will happen in the patent office is often worse than dealing with the granted patent. It's clear that with no probability of a patent, financiers would likely not have invested their dollars, telephone technology would undoubtedly have been delayed, and there certainly would have been no AT&T.

This vicious cycle is what happens when the patent office refuses to grant patents and when investors are unwilling to step forward and fund new technologies. When the financial infrastructure shuts down, nobody wants to invent, and innovation is stifled. And that's what we have today.

———

The case for protecting the little guy, the struggling inventor, lost most of its traction over the invention of the automobile and the ensuing patent litigation where a single patent was asserted against nearly every automobile manufacturer, including Henry Ford. The patent system had been created to protect the independent inventor at a time when large corporations were a rarity. The irony of automobile patent suits was that it was a single inventor taking on large corporations that turned the tide against the patent system. In a strange twist, a patent model—the Brayton automobile model—ultimately won the case for Henry Ford. Yet it was during the same decade the patent model requirement was eliminated.

The paper patent practice started almost the day after the model requirement was eliminated. And one of the first cases nearly put the

Ford Motor Company out of business. Yes, it was a paper patent, used by a single individual who had never made an automobile, that brought the entire motor car industry to its knees, forcing nearly every company to pay a royalty. Only Henry Ford, standing alone and sacrificing all of his profits, was able to free his company from this patent.

Today, these individuals are often referred to as patent trolls because they have no underlying business and obtain paper patents just to extract royalties. It's difficult to even pen the admission, but this first patent troll was a patent attorney—George Baldwin Selden, holder of the famous Selden patent on a "gasoline road locomotive." The son of a judge, Selden dutifully followed in his father's legal footsteps and became a patent lawyer in Rochester, New York. His credits include filing the first patent application for George Eastman on a process for placing a coating on gelatine dry plates. Although maintaining his legal practice appeased his domineering father, Selden's real passion was tinkering, inventing ways to manufacture barrel hoops and typewriters. His inventing went beyond the ordinary and mundane. Selden was mesmerized with mechanized road travel. On frequent visits to the patent office, Selden toured the Old Curiosity Shop, poring over models and drawings of machinery having anything to do with road travel. He salivated over rubber tires and steam drive carriages, hoping to find a suitable power source for his own horseless carriage.

Selden made little progress on his new road vehicle until 1876, when thirty years old, something clicked. He visited the famed Philadelphia Centennial Exposition where some of the world's greatest inventions were on display—Bell's telephone, an electric dynamo, an ammonia compressor for refrigeration, Otis Brothers & Co.'s steam elevator, not to mention some of today's household staples like Heinz ketchup and Hires root beer. The Mecca of Selden's pilgrimage was Machinery Hall, an exhibit over a third of a mile long housing the best machinery that the Industrial Age had to offer—the Corliss steam engine with its 56-ton flywheel that soared to the hall's ceiling, electromagnetic generators, and engines from Otto and Langen. Here, Selden hoped for the spark of an idea, something to propel his new vehicle. What caught Selden's attention was Brayton's gasoline engine—a compact compression motor, much sleeker in design than the bulky Otto engine, and one small enough to power his vehicle. Upon his return, Selden began in earnest to work on a modified Brayton engine that he could incorporate into his road locomotive. For specifics, he ordered details of the Brayton engine from the Vienna Exhibition of 1873. After hours experimenting with this engine, he made his decision: His patent application would include his own modified Brayton engine. If he'd only waited a year, he might have learned about Otto's new four-cylinder engine, the design that ultimately found its way into the modern automobile. Selden wasn't the only one trying to invent a horseless carriage. Inventors in America and in Europe were feverishly trying to find their own solutions. As a seasoned patent attorney, Selden knew he couldn't wait. He wanted to be the first to file his patent application. So Selden filed his application, describing the use of a Brayton-type engine, and submitted a non-working patent model.

Although there were no working automobiles, with the widespread use of engines, the idea of a horseless carriage was percolating, both in the U.S. and in Europe. A hundred years before, in 1769, a Frenchman named Cugnot constructed a steam tractor. Various types of steam carriages followed, including one currently housed in the British museum in London. Then came the electric motor, followed by the first electric car built in Scotland in 1839. When Brayton's and Otto's gasoline powered engines entered the scene, they offered a viable alternative to propel a vehicle—with significantly more power. To Selden's credit, nobody had yet invented a practical way to put a high-powered engine in a horseless carriage and make it work. And like any good patent attorney, Selden soon realized that it's much easier to conjure up a set of drawings than to actually build a motorized vehicle. With lack of funding to finish his motor and horseless carriage, and with his own patent litigation practice keeping him out of the shop, Selden knew valuable time was slipping away. He had to file his patent application, regardless of whether he could build a working vehicle.

And that is what he did.

In 1878, Selden went to work on his patent application. It didn't matter that his engine was still unfinished, or that he didn't really understand how to put a gasoline engine into a horseless carriage; under the new patent rules issued just a year before, he could give it his best guess. Selden took existing ideas—a doctored-up Brayton engine, clutch, transmission, and steering wheel—and hurried to see a patent draftsman to prepare a set of drawings for his new road vehicle. When the draftsman suggested it might be a good idea to submit a model, Selden told him that the drawings would suffice. Why not? Just a year before, the rule requiring a model submission was eliminated. And this created perhaps the biggest thorn in the side of the U.S. patent system—the paper patent.

The beauty of Selden's legal strategy was that it didn't matter whether he could make this road vehicle work. What Selden did know was that someday, somebody would figure it out. And Selden, the savvy patent attorney, knew he could manipulate the patent office rules so that he could keep his application pending until that day came. He could then change his claims to cover the yet-to-be-discovered horseless carriage. And so, with a little creativity, and lots of vague language, the world's first horseless carriage was created on paper and lodged with the patent office in 1879.

Falling squarely within permitted rules, Selden's legal maneuvering enabled him to keep his patent application secretly pending for another sixteen years, patiently waiting as the gasoline automobile became a reality. He was like a lion awaiting his prey. By 1895, when America's automobile industry burgeoned, Selden sprang to action, changing his original claims to cover essentially every vehicle with an internal combustion engine that used liquid hydrocarbon fuel. Because of how the patent laws were written, when Selden received his U.S. Patent No. 549,160, it had a future life of seventeen years. Remarkably, Selden had manipulated the patent system so that the time from filing his application to its expiration date amounted to nearly thirty-four years.

By the time Selden's patent issued, the automotive industry was beyond a fledging start-up. It was a legitimate industry. Daimler and Benz of Germany developed their motors in 1885. On the U.S. front, Olds and Ford began their ventures in 1894 and 1895. Dozens of other car companies sprang into existence—Buick, General Motors, Reo, Maxwell-Briscoe, Winton. It was like our later dot-com explosion.

Immediately, Selden went on a legal rampage, targeting the unsuspecting automobile companies. After joining forces with The Electric Vehicle Company, who took an exclusive license, Selden began enforcing his patent in the courts. At first, the loose association of automobile manufacturers thought they could beat Selden's patent, assuming that Selden's manipulation of the patent office rules would invalidate the patent. If that didn't work, they planned to argue that Selden's idea was not inventive—as he had just combined old, well-known elements. That strategy proved fatal for the Winton Company and its close ally, the Hydrocarbon Motor Vehicle Manufacturer's Association, whose request for dismissal of their patent infringement suit was denied by Judge Alfred Coxe in 1900. A month later, the same thing happened to the Buffalo Gasolene Motor Company. So, both cases headed for trial.

Bolstered by Judge Coxe's decision, The Electric Vehicle Company filed two more suits, one against the Automobile Forecarraige Company of New York and Ranlet Automobile Company of Vermont. Financially beleaguered, both companies capitulated. The gasoline automobile producers then banded together, hoping their organized forces could beat the patent. But after intense negotiations, the manufactures decided the best course would be to sign up for a license, then join together as an industry and force everyone else to pay the same royalty. Concluding their meeting, the automakers formed an association, the Association of Licensed Automobile Manufacturers, or A.L.A.M., and agreed to pay 1.25% on each vehicle. Left alone, Winton gave up and also joined the A.L.A.M. Small brass plaques began springing up on all automobiles with the inscription "Manufactured Under Selden Patent."

With twenty-three members, the A.L.A.M. went to work, systematically suing the other start-up automakers. Most, lacking sufficient resources, capitulated. Of the major manufacturers, the sole holdout was the Ford Motor Company. Unlike his competitors, Henry Ford wouldn't back down. Singlehandedly, he took on Selden and the A.L.A.M., funding his legal battle with profits from the sales of his vehicles. Nine years of litigation followed.

Not until January 1911 did the decision come down, and it was in favor of Ford. The court held that Ford's vehicles did not infringe the Selden patent. In essence, the court concluded that Selden's patent covered a gas-propelled vehicle with a Brayton-type engine, not the Otto engine used by Ford. Selden, the court ruled, described only a Brayton-type engine in his patent application.

In a sense, Ford's victory turned out to be hollow. By 1911, the Selden patent was ready to expire. Selden had already made his money, receiving royalties from nearly the entire automotive industry. Losing the patent at the end of its natural existence had no financial significance.

If the patent office had required models for just a year longer, none of this would have happened—no lawsuits, no attorneys fees, just more inventing. There simply would have been no Selden patent because Selden hadn't actually built a working prototype. Instead of spending valuable research and development money fending off lawsuits, the motor companies could have made better, more efficient vehicles much faster, perhaps even a decade sooner.

Where else has this happened? Look at Alexander Bell, who was not required to submit a model despite patent office rules calling for such models. If the patent office had stuck to its guns, Bell would *not* have been awarded his famous telephone patent and the giant Bell monopoly would never have existed. How different might the world be today if free competition could have reigned in the telephone industry from its infancy?

Is this same thing happening today? On a massive scale. By granting patents on concepts that have yet to be physically developed, our patent system allows technologies to be "locked-up", stunting development by at least a decade. Imagine where our electronics industry would be if we could jump ahead ten years.

This isn't an exaggeration. Beginning in the 1950s, Jerome Lemelson began filing applications on scanning visual data from a camera and storing them in a computer. These cases were kept alive in the patent office for a half century. When the barcode was eventually developed, Lemelson claimed his "machine vision" paper patent covered every single barcode. His litigation team sued every major company who made or used bar codes, netting him over $1.5 billion in royalties. It wasn't until 2004, well after Lemelson was dead, that a court threw out the patent on grounds that Lemelson had waited too long to enforce it.

Lemelson is legendary in patent circles. He's even got his own wing in the Smithsonian. Awarded over 550 patents by the U.S. patent office, Lemelson knew how to manipulate the patent system by flooding the patent office with applications, most never actually built. If he could dream it up, off shot an application, usually prepared by himself. His ideas included automated warehouses, industrial robots, cordless telephones, fax machines, videocassette recorders, camcorders, and the magnetic tape drive used in Sony's Walkman tape players. He also invented some medical instruments, including a talking thermometer for the blind. Most of these were mere ideas, patent applications filed in hopes that, as with Selden, the market would eventually solve the practical problems for him and in so doing, step into his waiting trap.

As with Selden and Lemelson, a large number of today's patent trolls would be put out of business if they were required to build their ideas before filing for patent protection. By preventing these innovation blockers from getting patents, valuable resources spent fighting dubious patent battles could be used to invent real products.

Chapter Eighteen

THE PATENT OFFICE CONFRONTS THE ETHER

For more than fifty years, Ellsworth's patent system had been meeting the needs of the nation's inventors, issuing more than 200,000 patents. With the exception of the eradicated model requirement, the process was mostly intact. And the best news was that patents were still issuing in a matter of months.

Then technology began to change, and the patent office had to come to grips with how it would handle scientific challenges, especially with no more patent models to help explain obscure ideas. Nowhere was that more evident than with the ether.

The concept of the ether had been around since Aristotle. It was the hypothetical substance through which electromagnetic waves, such as light, traveled. How could somebody patent a device that claimed to transmit signals through this mysterious medium?

The issue came up with Marconi's patent application for the world's first wireless device, the precursor to the radio and today's cell phones. Although today we think nothing of picking up a cell phone and calling a friend half way around the world, at the turn of the nineteenth century, if you wanted to communicate long distance, you had to do so over a wire, using Morse's dashes and dots. European and American cities were cluttered with miles and miles of unsightly telegraph wires, begging for someone to invent a more efficient way to communicate. Marconi was the person who sought to change that, making it possible to send a signal using Morse code, not over a wire, but through the "ether."

What Marconi claimed as his idea wasn't a way to transmit radio waves—that had already been demonstrated by Heinrich Hertz and James Maxwell. The real question was how to collect them. Marconi's patent claim was to a receiver that could take the radio signals and convert them to electrical pulses. He did this by putting metal shavings inside a tube and when the radio waves entered the tube, they oscillated the metal shavings, closing an electrical circuit. This Marconi did by developing a "coherer," a tube filled with iron shavings that conducted the radio waves. Marconi used the invention as a "wireless" substitute for the telegraph. For the first time, written messages could be transmitted through the air, eliminating the need to lay cables. His famous test case was the sending of the letter "s" from Poldhu, Cornwall, to St. John's, Newfoundland, in 1901, a distance of 2,000 miles.

One of the most practical uses of Marconi's wireless invention was on ships, previously out of communication with the shore as soon as they left it, with the British navy being the first to jump at the opportunity. These wireless devices became famous when used to capture the famous murderer, Hawley H. Crippen, and his mistress after the captain received a radio message apprising him of Crippen's crime. Then in 1912 his radio was used in the rescue efforts for the *Titanic*.

Marconi filed his patent application on December 7, 1896. His patent application consisted of 27 pages of description—all in cursive writing—with eight figures showing both mechanical drawings and electrical schematics of his receiver. By January 12, 1897—a mere month later—the examiner had reviewed Marconi's application and provided Marconi with his comments in a communication called an "office action." The comments mailed from the examiner reveal that the application was thoroughly examined. For example, the examiner agreed that the idea was new, but had several concerns about how Marconi had described his idea and, more importantly, about how it worked. The examiner's concerns were based on the prevailing belief that sending long distance signals through the ether was impossible.

Before granting the patent, the examiner wanted some clarification about how Marconi's radio worked. The examiner stated that, "On page 14 it should be explained how the receiver and transmitter can co-operate unless the two instruments are in view of each other." From this, it appears that the examiner believed that wireless communication was impossible unless the transmitter and the receiver were within eye sight of each other. If otherwise, the examiner wanted an explanation. The concept of transmitting through the "ether," although widely discussed, was thought to be an impossibility.

By May of that same year, Marconi filed his response—bold and straight to the point. "In reference to the Examiner's statement that the operation of the receiver and transmitter as contained on page 14 of the specification should be explained, it is submitted that this would involve a knowledge of the form and method of transmission of electric waves and appears to be beyond present scientific knowledge. The fact is known that in such a combination as disclosed by the applicant, the receiver is affected by the transmitter and signals can be sent from the one station to the other. We submit that it is immaterial, whether the path of the waves is in a straight line from transmitter to receiver through intervening objects, or whether the path is otherwise."

If a model was still required, the examiner could have asked Marconi to prove that the receiver really did receive a signal. Instead, the examiner took Marconi at his word. The examiner's actions are similar to what happens in today's patent office where a mere statement by the inventor's attorney is usually enough to have a patent granted. In Marconi's case, he really did have a functioning device and rightly deserved the patent. But that is not always the case, especially with many of today's patent applicants.

With Marconi's additional changes to the description to correct some clerical errors, the examiner allowed the case. In July 1897, a mere seven months after filing, Marconi had his patent in hand. Life could have been much easier for both Marconi and the examiner if Marconi had submitted a model, or some kind of demonstration proving how his idea worked. If Marconi had submitted a patent model, the examiner would probably have allowed the application without making any rejections. He would not have had to question Marconi about how it worked, thereby lopping months off the time his application sat in the patent office.

The seven months that it took to examine and approve Marconi's application, even when claiming a significant, ground-breaking invention, was not unusual. Although examiners wanted proof that an idea actually worked, they were still able to move the applications through the patent office with a rapidity that today is impossible.

Two decades before Marconi, Thomas Edison's patent application on the phonograph followed a similar path. It too relied on transmitting signals through air. But unlike Marconi's radio, it was the human ear that did the receiving.

Edison filed for patent protection on his "Phonograph or Speaking Machine" on Christmas Eve, 1877. The basic idea behind Edison's phonograph was to allow a person to speak into a mouthpiece and then record the sound waves on a cylindrical drum covered with tin foil. Although others had recorded sound decades earlier, like Leon Scott de Martinville's phonauthograph of 1855, Edison claimed to be the first person to play back the recorded sound. Edison stumbled onto the idea while attempting to record readable traces of a Morse code signal onto a disk. Edison proved his technology worked by singing "Mary had a little lamb" into the mouthpiece.

As with Marconi, the patent examiner took up Edison's case a mere month after filing, rendering his initial opinion on patentability. In the office action, the examiner was concerned that Edison had failed to sufficiently describe parts of his invention, stating that "the means for reproducing the recorded sound vibrations referred to on page 21 should be fully described and shown in the drawing in order that no question may arise in the treatment of any future application as to the sufficiency of such bare description unaccompanied with drawing." Another issue was that some of the material in the application overlapped with some of Edison's other applications, and the examiner asked for that description be stricken. The final issue was that, "No model has been filed showing the features covered by claims 5 and 6." The good news for Edison was the examiner's conclusion: "Upon suitable amendment the application will be allowed."

Even though Edison had established himself as one of America's greatest inventors, the examiner insisted that Edison clarify how he made his invention work. But as soon as that was done, the examiner was willing to allow the patent. It was that simple.

Edison readily complied with the bulk of the examiner's requests. A month later, on February 4, 1878, Edison's New York attorney, Lemuel Serrell, filed a reply with the patent office where he clarified the materials he used for recording sound, stating that: "The material employed for this purpose may be soft paper saturated or coated with paraffine or similar material with a sheet of metal-foil on the surface thereof to receive the impression from the indenting joint." He also erased any description of concepts covered in other Edison applications and canceled several claims because no model was submitted in connection with those claims. But Serrell didn't give in on every point. Serrell disagreed with the examiner about the insufficient description of how the device recorded the vibrations. "It is believed that the reference in the official letter of 26th Jan., to the insufficiency of the description in paragraph 21, is an error as the devices referred to are shown in Figs. 3 and 4." This argument was persuasive and by February 19, 1878, a little over a year after filing his application, Edison had patent number 200,521 in hand.

Patents in less than a year—all without computers, electronic filings, or the Internet. And, all while struggling with significant advances in technology. Back then, the patent office was willing to grant patents on leading edge technologies, even if the advances were small. Today, all this is impossible. Besides the lack of efficiency in today's patent office, the patent system has layered on thousands of additional requirements over the years, making the patenting process onerous at best. In large part, this has *not* been due to overly complex technologies, but from a fear that patents are too powerful. The result is that it now takes three to five years to get a patent application through the patent office, even on the most simple of technologies. Even then, the claims are usually narrowed so far that the resulting patent is often worthless.

Issuing patents on ideas that delved into the ether was one thing, but what would happen if someone patented a flying machine? The patent office had a strict rule about flying machines—they were not patentable. Would the patent office be willing to admit that it was possible to fly? Even more important, if someone really could fly and had the only patent on how to do it … That just wouldn't be right. That would mean one patent could control what flew in America's skies for more than a decade. But that's exactly what happened.

Chapter Nineteen

IN COME THE WRIGHT BROTHERS

If Eli Whitney was on the leading edge of America's innovation boom, Wilbur and Orville Wright came at the end. Between them, the amount of innovation is unparalleled. For the Wright Brothers, the U.S. patent system was both friend and foe: foe because of the ban on granting patents claiming flying machines, and friend because the patent, when it was finally granted, gave the Wright Brothers a virtual monopoly over any airplane. And what happened to the Wright Brothers foreshadowed what was to come to future generations of inventors.

The biggest problem faced by the early aviation pioneers turned out to be quite different than anyone expected. To most, it was simply how to keep a body from falling out of the air while propelling it forward. But by the turn of the twentieth century, experimenters had discovered that it was quite possible to glide a machine through the air on wings. The real problem was how to keep the wings from tipping from side to side. Controlling the plane when turning was even more difficult. Without the necessary controls, the flying machine would simply crash to the ground. This was where the Wright Brothers focused their efforts.

Wilbur and Orville Wright grew up under the hand of an evangelical father who found himself entangled in legal battles over church property after his congregation split. Wilbur, an obsessive reader, assisted his father with his legal issues. The skills he developed thinking through legal issues would help Wilbur in solving the problem of how to keep a flying machine in the air. Orville was the first one to take on a business, starting a printing press as a teenager. In 1893 Wilbur opened a bike shop, selling a new kind of bicycle with two wheels of the same size. Orville joined his brother, although Orville's dream was to start an automotive company. Wilbur suggested it would be easier to make a flying machine than to make money building cars.

When Orville Wright became deathly ill with typhoid in 1896, Wilbur made a vow that if his brother was spared, together they would dedicate their lives to solving the problem that killed "The Flying Man," Otto Lilienthal, during his attempt at flight. So when Orville returned to health, Wilbur immediately ordered a library of books, pamphlets, and publications on aerodynamics. Wilbur quickly discovered that their biggest problem would be finding a way to control the lateral stability of the flying machine. Traditional thought was to keep the wings level, even while turning. Wilbur's genius was in taking an opposite approach and letting the plane roll during turns, in the same way he banked his bicycle during turns. Identifying how to do this with an airplane became his quest.

The Wright Brothers solved the problem of lateral stability by warping the wings, which involved bending the edges of the wings at their far ends. The idea came to Wilbur in July 1899 when throwing an empty bicycle tube box into the garbage. Noticing the twisting box sparked an idea. When the airplane began tipping to one side, the plane might be stabilized by twisting or warping the wings in a certain way. Starting with kites and then graduating to gliders, Wilbur perfected his wing warping idea. Next came prototypes, then Kitty Hawk. It wasn't until March 23, 1903 that the idea was perfected enough so that Wilbur felt comfortable in filing his patent application. Today, ailerons are used to stabilize airplanes, but the original concept of how to stabilize an airplane came from Wilbur's wing warping idea.

The Wright Brothers' experience with the patent office was different than that of most of their peers. In their first attempt, their application was rejected outright by the patent office, who refused to grant any patents on flying machines. It took a full three years, until May 22, 1906, for the Wright Brothers to convince the patent office to issue their patent to the airplane.

What few realize is that the battle over who owned the patent rights to the airplane, and thus the ability to decide who could fly in America's skies, came down to how to control an airplane in flight. In other words, the Wright Brothers didn't claim to own the general principle of flying, just one specific way to practically accomplish this. But it turned out that nobody could figure out any other way to keep a flying machine in the air. In essence, then, the Wright Brothers patent did control the skies. In turn, that generated the controversy over their patent.

While the Wright Brothers began pursuing their patent in 1903, their only funding (a meager $1,000) came from their bicycle shop. After their patent issued, they were unable to find interested investors in America, so they traveled to Europe and signed a contract with France. Back in the U.S., those interested in the aviation business chose to produce their own airplanes, rather than finance the Wright Brothers' business. The most successful of these early manufacturers was Glenn Curtiss.

With an uncooperative patent office that was unwilling to issue the Wright Brothers' their patent, the Wright Brothers were left without recourse, helplessly watching as their patent languished in the patent office and competitors slipped in. Even after issuance, the Wright Brothers still faced the issue of patent enforcement. They could barely finance their own operations, let alone take on an entire industry—even if they had created it.

The irony of the Wright Brothers situation was that their patent didn't become valuable until their competitors became successful and the airplane business took off in the U.S. As soon as the world recognized the airplane for what it really was, they also realized how powerful the Wright Brothers' patent had become.

Seeing the Wright Brothers' patent as a revenue source, American businessmen organized forces by forming the Wright Company, issuing $1 million in stock with a paid-in value of $200,000. These businessmen believed they could license the patent and collect large royalties. But to do this, they first needed to stop Curtiss from selling infringing airplanes. Investors included Cornelius Vanderbilt, August Belmont, and Robert Collier, with J.P. Morgan withdrawing his investment for fear that he would overshadow the other investors.

Those who assumed the Wright Brothers would not obtain their patent and took the risk of selling airplanes during the three years while the patent application was pending in the patent office were now facing a serious problem. They were now infringing the Wright Brothers' patent. The infringers tried to fend off the inevitable. In 1908, the Aero Club of America sought a solution to the infringement issue by attempting to raise $100,000 to purchase U.S. rights to the Wright Brothers' patent. This venture failed, but Curtiss and his Herring-Curtiss Company didn't stop making airplanes. This forced the Wright Brothers' hand, and the Wright Company sued Curtiss for patent infringement in 1909.

The Wright Brothers were fortunate in that the judge handling their case against Curtiss was the same judge who handled Selden's first automobile infringement case. Judge Hazel found for the Wrights in January 1910 and issued a preliminary injunction. Curtiss appealed and six months later the injunction was lifted after Curtiss posted a $10,000 bond. The rest of the infringement case would drag on for seven more years until World War I prompted government action to settle the case.

Still, from 1909 to the start of World War I, the Wright Brothers' patent effectively shut down all competition in the U.S. while they awaited the outcome of the case against Curtiss. The patent was generally viewed to be so strong that in 1915, Eastern investors bought the Wright Company from Orville for $1.5 million.

The decade-long battle with Curtiss may have had the greatest psychological impact on turning the courts and the public against the patent system. The Wright Brothers claimed that their patent covered any way to laterally control the plane in flight, referred to as the roll of the plane. To do this, the Wright Brothers used wing warping, a way to twist the plane wings at their leading and trailing edges by using a system of cables to control the up-and-down movement of the wing tips. The other way to control the roll of the plane is to use ailerons. Herein lay the problem—the Wright Brothers claimed their patent covered both ways, even though the patent described only wing warping. And the Wright Brothers were correct if you read the literal language of the claim, which states that the wing has lateral marginal portions that may be moved to different angles relative to the normal plane of the wing. Both wing warping and ailerons involve moving parts of the wing off a normal plane.

But the public didn't view it that way, especially when courts used language stating that ailerons were "equivalents." The decade-long case had tested the public's patience. Many felt Orville Wright was overreaching when he claimed his patent covered any way to laterally control the plane in flight even though the patent described only one way—wing warping. Ironically, it was Alexander Bell—the innovator accused of stealing broad language describing the telephone from Elisha Gray's patent application— who spearheaded a charge to limit the scope of the Wright Brothers patent. And he did it by using Ford's former patent attorney, who continued to fuel the fire by encouraging Curtiss to sell airplanes with ailerons, rather than using wing warping. The end effect was to create a debate over patents that would last for a century. As pro- and anti-patent forces fought for turf, the patent system became more complex and unruly until what happened in the Ford and Wright Brothers cases seems trivial.

The concept that a competitor could infringe a patent, not for doing exactly what the patent claimed, but doing something "equivalent" to what it claimed caught hold in American jurisprudence. The effect was to increase the scope of patent claims, in essence making the patent broader in scope, and a more powerful weapon. It's no wonder that courts tried to curb this power. But how they did it only served to complicate the patent landscape, increasing the cost to obtain, enforce, and defend a patent. Rather than simply eliminating the doctrine of equivalents to reduce the power of a patent's reach, courts and the legislature chose to make it harder to obtain a patent (or easier to invalidate an existing patent). The principal way to accomplish this was—and still is—to permit a patent to be denied or invalidated if the invention is "obvious," and this has made it nearly impossible for small inventors to participate in the protections afforded by the patent system.

Chapter Twenty

WHERE DID THE INVENTORS GO?

Where are the likes of Goodyear, McCormick, and the Wright Brothers today? The truth is that they are hard to find. The vast majority of today's commercial products come from large companies—who aren't inventing. Major corporations aren't set up to innovate. They are fantastic at marketing, complying with regulations, and keeping products safe, but they are not focused on inventing. As things stand, the successes of nineteenth century are unlikely to be repeated.

The question is why?

After the decade of the Wright Brothers, inventors fled to the newly created research labs for secure jobs with safe retirement plans. The ranks of Bell Labs and IBM swelled. It wasn't until the 1960s and 1970s that the computer revolution took hold and innovation again began to spark. But although they have certainly changed the world, today's computer moguls—Steve Wozniak, Bill Gates, and Paul Allen—traveled paths that are fundamentally different from those of their peers during the Industrial Revolution. Patents played almost no role in protecting the first computer operating systems. Today's potential innovators have little incentive to develop products that cannot be reasonably protected with a patent. The patent system is too unfavorable, and few individuals succeed. Perhaps the last of these warriors was Bob Kearns, inventor of the intermittent windshield wiper. But it took his entire life and years in court to obtain vindication. Today, Kearns wouldn't stand a chance.

What has caused the decline in innovation, and more importantly, the dampening of the inventive spirit? Although this country gives lip service to helping out the little guy, the truth is that those in power don't want to deal with the little guy. We don't want the thousands of farmers and mechanics improving the reaper and getting patents.

Why?

If they did, our complex court system would bring industries to a screeching halt—just as it did to the airplane business, but worse. Much worse.

Anticipating that consequence, judges and politicians watching the Wright Brothers' battle were fearful of what patents could do to an entire industry. They had to stop patents from issuing too easily, from letting too many patents pass through the patent office. The courts now wanted inventors to show that they had made significant, non-trivial advancements over the current state of technology.

So today we give patents to only those ideas that are so-called "truly inventive," not just new. In other words, coming up with a new idea doesn't get you a patent. The invention has to be above and beyond an "ordinary" invention to get the patent office stamp of approval. Put another way, the invention can't be "obvious" and not inventive enough for patent protection.

How does this work?

Logistically, examiners can declare a claim "obvious" by taking bits and pieces from different documents, putting them together, and saying that it would have been obvious to combine all these pieces to get the invention. The problem with this approach is that every invention is cobbled together from the scraps of old ideas. Therefore every patent application can be rejected if the examiner simply declares the idea to be "obvious."

Take a simple example. Suppose you're Steve Martin's character, Navin Johnson, in *The Jerk*. In the movie, Navin invents a pair of slip-proof eyeglasses with a lifter on the bridge, which he calls the Opti-Grab device.

A replica of these glasses looks like this:

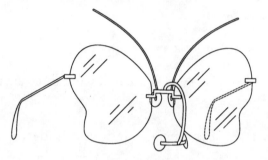

Aside from the fact that the glasses make the wearer go cross-eyed, assume that Navin has a marketable idea he wants to patent. Navin brings his prototype to his patent attorney with this drawing. It looks something like this:

Navin explains that his glasses are unique because they have a handle that lets the user readjust the position of the frames without touching the lenses. Also, the handle has nose pads that sit near the wearer's nostrils, "grabbing" the nose to prevent the glasses from slipping. That's why he calls it the Opti-Grab.

Using this description, his patent attorney drafts the application, demonstrating how normal glasses slip down the bridge of the nose and how this new design, with a handle jutting from the frame, will grip the nose and allow the user to readjust the position of the glasses with ease.

The attorney describes the invention in detail, adding reference numbers to the drawings so that when the examiner reads the patent application, she can look at the numbers and locate them on the drawing. Then comes the hard part—the claims. This is all that really matters, because after a patent claim issues, it tells the world what others are excluded from making.

A simple claim might look something like:

1.   A pair of sunglasses with the following three elements:

    a) A frame;

    b) A pair of lenses separated by a nose bridge;

    c) A lifter, connected to the nose bridge, which the wearer can grab to adjust the position of the frame.

When the examiner reads the claim, his best grounds for rejection will be to find a single document, referred to as a piece of "prior art," published before Navin filed his patent application and that has all three elements of the claim: a frame, a pair of lenses, and a lifter on the nose bridge. This piece of prior art can be an issued patent, an article, a newspaper advertisement, and so on, but it must have all three elements of the claim. If it doesn't, then it is not an "anticipatory" reference. In the days of Colt, Goodyear, and McCormick, this was the only way claims could be substantively rejected.

But today, the examiner can resort to another type of rejection: that the claim is "obvious."

This notion of an invention being "obvious" and therefore not entitled to patent protection came into being around the time of the famous patent trials of the 1850s. Evidently, the Supreme Court didn't like the patent office giving patents away so easily and thought there had to be more blood, sweat, and tears, or some type of "flash of genius" before granting a multi-year monopoly. When this doctrine was first articulated, most courts ignored it. Over the next century, however, as the concern over unwieldy patents continued to mount, more courts took notice and began invalidating patents based on various forms of the obviousness doctrine. In an attempt to clarify the doctrine of obviousness, Congress passed a statue in 1952, codifying how patents could be rejected as being obvious as 35 U.S.C. 103.

The concern of the unfriendly patent crowd was that an inventor like Navin could just take a well-known thing like eyeglass frames and combine it with something simple, like the handle on a coffee cup, and land a patent that prevented the rest of the world from using glasses with a lifter for the next few decades.

The problem is that *every* invention is a combination of known elements. The statute makes an examiner's job of rejecting patent applications easy: just cobble together a few patents that, when combined, have all the elements of the claim and then say that an invention is obvious. And that's exactly what the patent office—with the blessing of the Supreme Court—does in probably more than ninety-five percent of patent application cases. Here's how a typical rejection is formulated:

First, the examiner searches in vain for one single reference to a frame, lenses, and a lifter but can't find one because Navin really did come up with a novel idea. But the examiner has at his disposal a thousand prior patents showing various kinds of eyeglasses. So he picks one of them and says he's found a reference with a frame and lenses.

Now the examiner finds a second reference showing some kind of handle. It doesn't really matter what kind of handle; any kind that can be used to lift things. The examiner wades through all the handle references and picks out one pertaining to a coffee cup, because this vaguely resembles the handle in Navin's drawing.

Now he's got Navin with a perfect obviousness rejection. He drafts a sentence that says, in essence: Navin, your eyeglasses with a lifter is obvious, because "one of ordinary skill in the art" would have taken an ordinary pair of eyeglasses and modified them to include the handle of a coffee cup. As such, your claim is obvious and stands rejected.

Of course, Navin's attorney can argue that this is utter nonsense, but he might as well save his breath. After an examiner takes the position that a claim is obvious, it is next to impossible to change her mind that it's not obvious. Most often, the patent isn't granted, except for those with enough money and persistence to manipulate the system until the examiner, worn down, and feeling that the patent office has collected enough fees, grants the patent. Naturally, this results in a complete lack of certainty as to who gets a patent and who doesn't. And with the current hostility toward inventors in the patent office, not even time and money seem to be working.

In the 150 years since the obviousness standard was created, the big debate has been to determine exactly what is obvious and what test should be used as the standard. For example, in the 1941 case of *Cuno Engineering v. Automatic Devices Corp.*, the Supreme Court came up with the harebrained idea that to be patentable an invention must "reveal the flash of creative genius, not merely the skill of the calling." What is a flash of genius? Should Navin submit a journal entry at the time he invented his lifter that he was in a state of nirvana? Even Navin has enough gray matter to tell this is a ludicrous test.

Following the 1952 patent act, the courts removed the "flash of genius" phrase, and several tests have come in and out of fashion ever since. The latest pronouncement by the Supreme Court came in 2008, when they threw down the gauntlet. They evidently thought patents were getting out of hand, most likely because of an earlier appellate court decision by the patent appeals court, which permitted the patenting of software patents. In 1998, the Court of Appeals for the Federal Circuit in *State Street Bank v. Signature Financial* decided that software and so-called business methods could be patented as long as they were new and non-obvious, as with every other technology. Then came the Amazon one-click patent, the bad press that followed, and the patent office's moratorium on issuing software patents.

Most people don't know this, but it's true. The patent office actually stopped issuing patents in several technology areas. Just ask any examiner or patent attorney who handles software patents, and they will tell you: one day the director was getting a lot of heat from the press over the Amazon patent, so he told all the examiners to do whatever it took, just don't issue any more patents. And they didn't—and we're still reeling from the effects of this action.

Of course, the patent office tried to be discrete about the mandate, but it ended up being a giant snafu. When an examiner picked up a case, he would simply pick any reference to reject the claim. Literally, it could be any reference, it didn't matter what it was, relevant or not. He'd just say your application is obvious in view of such and such a reference.

Evidently, the Supreme Court agreed with this practice and in its 2008 *KSR* decision placed their stamp of approval on these type of "obviousness" rejections. They established yet another test, now referred to by patent practitioners as the KSR test. It basically states that if the examiner thinks it's obvious, then it is. In legalese, the court's opinion states: "If a person of ordinary skill in the art can implement a predictable variation ... [it] is obvious." But what is predictable?

There's more from the Court: "When a patent simply arranges old elements with each performing the same function it had been known to perform and yields no more than one would expect from such an arrangement, the combination is obvious."

Translation? If an examiner says your invention is obvious, then you're patent application is going to get rejected. And it will remain so unless you have enough time and money to wear down the examiner.

How much? Navin would have a heart attack when getting the bill from his attorney. Arguing three or four rounds with an examiner can cost around $10,000 to $12,000, on top of the $15,000 he already paid his attorney to file the application. And four years will have gone by since he filed his application.

This hypothetical isn't an exaggeration. A few years ago I was engaged to secure the patent for a new and upcoming shoe company, one that made a "squishy" sandal that looked like a clog but had the trademark of a crocodile. That was in the days when this little startup lived on a shoestring budget—literally—making their first payments in shoes instead of cash.

When their patent application was examined, it too received its share of "obviousness" rejections. I made multiple trips to the patent office, toting samples of the shoe into the examiner in an attempt to show why this new sandal was deserving of a patent.

The stakes were high. Millions of knock-offs were flooding the market, and I couldn't seem to get this patent application through the patent office.

"These are kind of cute," she said, inspecting a light blue shoe.

"These are the real ones you've heard about. A lot of the other stuff you see on the streets, those are knock-offs." I handed her one of the knock-offs. "Take a good look at them. You can't tell a difference. They look identical, and for all I know came out of the same factory. You know China..."

She nodded in agreement.

"So let's cut to the chase. I've got to get a patent. Any patent. It doesn't need to be broad enough to cover the world of foam clogs, just enough to cover this one." I pointed to her light blue sandals which she kept passing between her hands, squishing the pliant foam. "You can see that these guys are unscrupulous. They didn't even try to make their clogs look different. They are killing my client, an all-American company, right in the heart of Boulder, Colorado. If I don't get this patent, the Chinese are going to come in and clean house."

She paused, looked at me, and lowered the shoe. I knew what was going through her mind, because she'd told me during my last visit. She didn't approve of lavish copying of an invention, but if she allowed a patent application that ended up in litigation and the press made it look like a silly patent, she'd catch it. "Look, I want to help you," she said, "but you know what it's like around here."

This time I nodded. She didn't need to elaborate. Everyone knew the story. The patent office was reluctant to issue patents. In private, many patent examiners referred to the patent office as the "non-patent office."

"I understand. That's why I'm here. Let's see if we can't work something out that will get past your boss, the second set of eyes, and all the other quality assurance people who don't want this to issue. Let's narrow the claim to where you feel comfortable, then you can allow the patent, get your points, and we can both be happy."

She bit her lip and her eternal smile fell.

"What?"

"I'm still having a hard time with this one. It's … it's …. well, it's just obvious."

"Come on," I said. "How many references are you throwing together? Look, you've got a wooden clog, a patent on some foam, and a woman's pump with a heel strap. You're throwing them all together and saying that's my client's shoe." I held my breath, waiting for my moment of frustration to pass. She was just doing what she'd been told to do—rejecting patent applications for being obvious.

The patent on this new sandal did eventually issue—and it did end up in litigation where the court ultimately upheld its validity. But it cost a fortune. It was only because of a successful IPO that this shoe company was able to wage the war. Most others aren't as fortunate.

And so large corporations—those with enough money to fight the battle—manage to get patents while others do not. So in trying to fix the patent problem by reining in patents, Congress and the courts have stifled innovation among the very class of inventors that we want to be inventing.

The end result is that today we award patents to only the wealthiest of corporations with the financial means to battle with the patent office. This upends our Constitution which directs Congress to promote innovation. But with the obviousness doctrine, innovation is not being protected, only the illusive concept of "inventiveness." And, the only ones who can prove that their ideas are "inventive" are the ones with endless funds in their coffers. The end result is that the vast majority of potential innovators don't invent. The doctrine of obviousness may be great for companies like IBM, who file 10,000 patent applications each year and collect hundreds of millions of dollars in royalties, but it kills the inventive spirit of everyone else. And history has proven that our best ideas come from the ordinary person. Ironically, the patent system has now become so unruly that even many of the large corporations are crying foul.

With the high-tech revolution and the introduction of the dot-com era, we stood the best chance of having another technology boom—one that produced successes similar to those of the 1830s. But in large part we failed. Yes, we came up with some important technologies, but not on nearly such a grand scale. And when it comes to the energy business, we've fallen flat.

What is disturbing is that today's patent rules are so onerous that they would invalidate most of America's greatest patents—the airplane, the telephone, the transistor, the light bulb, the electric motor, the phonograph, Marconi's wireless device, xerography, just to name a few.

Just look at the first patent on talking pictures—U.S. Patent No. 823,022. The application was filed in 1905 and covered a way to place sound grooves in film so that as the film was played to produce images, sound could be reproduced as well. Going to the movies is America's number one pastime, vastly more popular than baseball. Surely, today's patent office would deem this idea patentable.

Highly unlikely.

One thing is for certain: Today's patent office would be incapable of granting this patent in less than nine months, as it did in 1905. Today it would take a year or two to even assign it to an examiner.

Then there's the claim. It was about as simple as they come: a piece of film with pictures on it and a series of sound grooves used with a phonograph.

So what would happen if a patent examiner got hold of this application today? Showing silent "pictures" was well known in 1905 and Edison's phonograph technology had been around since 1878. Today's patent examiner would almost certainly hold that combining the two technologies would have been an obvious thing to do. It is all but certain that America's most popular invention would be deemed obvious.

And that is why the little guy in America has stopped filing patent applications.

Chapter Twenty-One

THE STAGGERING COST OF INVENTING

The cost to obtain a U.S. patent is overwhelming for most inventors. In 1850 an inventor could expect to pay around $50 to obtain a patent, a price that included the $30 government filing fee. Today that number is well above $25,000—a price that is mostly due to the problems in shepherding the application through the patent office.

Litigating a patent against an infringer is a whole different story. The outrageous cost of obtaining a patent pales in comparison with what it takes to enforce a patent. Although patent litigation has always been expensive, even during the boom times of the 1850s it is nothing compared to today.

The patent wars of the 1850s—the result patents being issued by the thousands in the 1830s and 1840s—were front-page news, in large part because of the personalities involved and the massive amounts of money invested in these cases. McCormick paid his lawyer Stanton $10,000, just to be topped by Goodyear, who paid Daniel Webster $15,000. A half century later, Henry Ford spent over $250,000 to prove that he did not infringe the industry-blocking Selden patent. These figures were widely circulated at the time, creating the impression that patent infringement cases were outlandishly expensive.

Today, the costs of even routine patent cases are in the millions. According a 2009 American Intellectual Property Law Association survey of the patent profession, costs to take a patent case to trial are based on the value of the damages. For cases valued over $25 million in damages, the data shows that it costs $3 million on average to get the case to the end of discovery and a total of $5.5 million to reach a jury decision. With these figures, how can anyone expect to enforce a patent—assuming they can get one in the first place?

It's a nearly insurmountable obstacle facing small inventors and corporations. Suppose a small company discovers an infringer who just took several of the company's accounts, selling them the infringing product at a lower price. The small company doesn't want to sue, mostly because they don't have the $5 million required to litigate. So what can this small company do? No start-up can afford to spend the money required to litigate and any sophisticated infringer knows that.

So the small company decides to sit down and talk with their competitor, in hopes that they will be reasonable and take a license. To get started, the start-up either calls the infringer or sends a letter explaining that they have a patent and they'd like to have a discussion about how to handle the infringing sales. Based on this action alone, the powerful corporation can run to court and sue the patent holder for what's called a declaratory judgment, an action to have the patent declared invalid. They do this in a part of the country that is favorable to them, where they can be sure it will cost the small company a fortune to litigate. Now the patent holder must spend millions just to save its patent from being invalidated. Of course, the patent holder can counter claim for patent infringement, but if the small company doesn't have a war chest of millions of dollars, they are only digging their grave with two shovels instead of one.

To avoid this, the small company must first file its own patent infringement suit in the jurisdiction they think is most favorable. They then call up the infringer, apologize for suing them while in the same breath explaining why they had to do it this way. Not a good way to start a settlement discussion. And, this course of action really isn't an option because it involves suing the infringer, which that small company can't afford.

Then what is a small company to do? Even with the odds stacked against America's innovators, some strategies can still work within the patent system, although none of them are ideal. Patent holders can use the court system to their benefit in cases where potential damages from infringing sales are high. They can do so through the use of contingency litigation—where the patent holder hires a law firm that is willing to take the case for a portion of the fee. There are literally thousands of ways these deals can be structured, but they each involve the law firm getting a piece of the action. This gives the law firm an incentive to win the case because they now have a vested interest.

Contingency litigation, however, involves immense risk. To take a major piece of patent litigation all the way to trial could cost a law firm $5 million to $10 million, or even more. Being on the losing end can put even large law firms out of business if not managed appropriately. So before a law firm will take patent litigation on a contingency fee arrangement, they will want to compensate themselves for taking the risk. And the law firm's take is based on the size of the damages award. For example, if a law firm receives one-third of the recovery, the law firm will want a case where it gets at least three times what it spends on the case. So, if the case requires the law firm to invest $5 million, the law firm will want to get paid $15 million if it wins. Because the law firm is getting a third of the recovery, the total damage award should be $45 million. Typically, damages are based on reasonable royalties, which often hover around 5%. To collect $45 million, that means there must by $900 million in infringing sales. Few products reach this amount of revenue generation.

Even so, patent holders can succeed and the stories are often heartwarming. Take the case Dr. James White, the college professor who invented a way for disk drives to read and write data without crashing. When Dr. White began his research, disk drives were constructed of a rotating disk, called a hard drive. The electronic component that recorded data onto the disk was called a read/write head. An arm held the head above the disk and moved it to the correct location to read or write the data. However, the head had to float above the disk surface by a set distance to record the data onto the disk and retrieve the stored information. The key to getting the whole thing to work was called a slider—the component that held the read/write head below the arm. As the disk started spinning, the slider would begin to float above the disk. Think of the slider as a miniature airplane flying above the disk surface. But if something changed the trajectory of the slider, usually when the disk was bumped, the slider would crash into the disk. Not only did this interfere with the recording of data, but it could ruin the disk.

Dr. White solved this problem by creating an aerodynamic design for the slider that kept it from crashing into the disk. He developed very elaborate computer models to simulate the flying conditions and kept changing the design until he found a stable slider configuration. Then he figured out a way to manufacture his design, using techniques from the semiconductor industry.

After being awarded his patent, Dr. White took his idea to the computer industry. They all loved the design—so much so that it ended up in the majority of disk drives. The problem was that nobody wanted to pay him for it. They stalled and stalled in their negotiations, knowing a college professor couldn't possibly have enough money to bring a patent infringement suit. That's when he came to my firm, where we decided to take the risk because the potential damages were massive. As a result, Dr. White sued most of the disk drive industry, most of which ended up taking licenses.

Dr. White's case is rare. The vast majority of America's inventors, or the people in America who should be inventing, aren't so lucky. Our most important class of inventors—the solo tinkerer or small technological start-up company just introducing a product to market—find themselves in a gaping abyss, abandoned by our legal system. They have an innovative product with good sales, perhaps in the range of $10 million in revenue a year. They are lean and have almost no room in their budgets for extra expenses. Because they have a niche market, total annual sales might peak at $50 million, so when the big corporation comes in and takes $10 million of the market, it really hurts. Yet no law firm is going to be willing to take this on contingency. There just aren't enough damages on the table. So here we have the bread and butter of America's idea pool who are unable to enforce a patent—if they can even afford to obtain one.

Is there no hope?

Perhaps some, though not much. To reach a settlement, the patent holder needs to find some other way to make an attractive proposal to the infringer and bundle the patent as part of this business arrangement. Of course, this won't give the patent holder the full value of his patent, but when there aren't many other options, it's one of the best. When going down this path, the patent holder must realize that flat out stopping an infringer is going to be tough. Offering a reasonable license in combination with other incentives is often a more realistic approach. And "reasonable" means a highly discounted royalty rate.

Perhaps the best example of this is T.J. Izzo, who invented the Izzo double-strap golf bag. Any serious golfer has seen this strap. Before its introduction, golfers (or caddies) toted the golf bag over one shoulder. With a heavy tour bag, this was worse than backpacking with a 70 pound pack, mainly because the weight was unevenly distributed on one shoulder. T.J. was a talented golfer and loved to hit the course whenever time permitted. He wouldn't take a cart because he loved to walk for the exercise. But when he hurt his back, that all stopped because the single-strap bag kept pulling his back out, taking a toll on his golf game. In 1990, T.J. came up with the idea of a dual strap golf bag. His first prototype was made from towels and duct tape. From there, he marketed the dual strap bag out of his home in Evergreen, Colorado. A year later, he formed IZZO Systems, Inc.

T.J. wasn't the first person to put two straps on a bag. That had been done at least a century before. What T.J. did invent was a way to use two straps while evenly distributing the weight and preventing the clubs from sliding out. If you've ever used one of T.J.'s bags you soon discover why they are so comfortable, enough so that nearly all the caddies on the PGA tour took to them instantly. The invention was so successful that Frank Thomas, technical director for the USGA, told Golfweek that "One of the four greatest innovations in modern golf is the double strap for golf bags, patented by T.J. Izzo." The rave reviews also made the *New York Times* after T.J. signed up the Wilson Sporting Goods Company of River Grove, Illinois, as its first licensee. "The new bag is designed to balance the weight of clubs more evenly across the back than single-strap bags do." The 1992 *Times* article stated. "'Once you see it, it kind of hits you in the head,' said Jim Grundberg, Wilson's business manager for golf bags. 'You can't believe the idea hasn't been around before.'"

After signing Wilson, T.J. found it tougher to make new deals. New licensees wanted the technology but didn't want to pay the license fee. Some were worried about already razor-thin margins. Others felt as if no one should get a patent on a double-strapped golf bag. Then there were those who thought they could design around the patent.

Rather than file lawsuits—which T.J. was smart enough to see would get him nowhere—he took a different approach. He'd call up a big bag manufacturer, like Nike or Titleist, and invite their executives to play a round of golf, followed by a nice dinner. Then he'd schedule a business meeting with them the next day. Of course they knew he was trying to sell them a license, but few people will turn down a free day of golf and dinner. When it came time to discuss the details of the deal, T.J. explained why the world of golf would take to his new bag and that they didn't want to be left behind. He offered them very reasonable royalty rates. But he also threw in other perks and incentives. He'd reduce his royalty rate even further if they'd let him do some advertising on their bags. His company also started sponsoring junior golf tournaments, which the golf equipment companies and the professional golf associations loved because it encouraged more people to golf. This generated immense goodwill and made it easy for manufacturers to want to do business with him.

Using this creative approach, T.J. was able to sign up most of the major equipment manufacturers before selling his company to well-funded investors who did have the money to enforce his patent—which they did. T.J. now dabbles in several other ventures, but mostly enjoys life while playing golf.

═══════════

The problem with patent litigation is even worse if you are small and on the receiving end of a patent infringement case—where large corporations can almost shut down a competitor at will. It takes just as much money to defend a patent infringement case as it does to bring one. Because of this, if a large company wants to force an emerging competitor to spend $3 million, all it has to do is sue for patent infringement. It can take a start-up to bankruptcy almost overnight.

Several years ago, I received a call from Maverick, a highly innovative mountain biking company. Paul Turner, Maverick's founder, was best known for inventing RockShox. He'd sold his interest in his previous company and was now marketing a new mountain bike frame. The new design had attracted many in the biking industry. The draw to this new frame was that it provided increased traction when going uphill, but was stiff enough on level ground to provide an efficient ride.

Maverick had filed for patent protection and was on the verge of joint development agreement with a leading mountain bike company. In an apparent effort to stall this deal, one of Maverick's well funded competitors, a well known American bicycle company, threatened Maverick with a patent infringement suit, then forwarded a copy of its cease and desist letter to the potential partner. The threat of suit for patent infringement had the desired effect, bringing Maverick's negotiations with the partner to a screeching halt. The potential partner backed out of the deal and terminated discussions. From Maverick's perspective, this was pure harassment—and an actionable intervention in Maverick's relationship with a potential joint venture partner.

The mood around Maverick was glum. On the verge of signing a major licensing deal, the company was now facing a multi-million dollar suit involving a patent that had almost no resemblance to Maverick's design. So what could they do? Spending millions litigating didn't make sense. But without a major partner, the technology would likely languish.

With few options, the company decided to take a gamble and went on the offensive. Maverick led off by filing a civil complaint against its competitor on January 10, 2001, in the United States District Court for the District of Colorado. In a nutshell, the complaint asked the court to declare the competitor patent invalid and to require the competing bicycle company to pay damages, attorneys' fees, and costs in connection with its willful and bad faith interference with Maverick's business relationship with the other bicycle company.

However, Maverick didn't serve the complaint, giving the parties about three months in which to discuss settlement options before the law suit would really begin. After filing the complaint, Maverick sent the competitor a letter informing them of the lawsuit, and clearly stating that the revival of the business relationship between Maverick and the potential manufacturer was critical if the matter was to be resolved amicably and without litigation. Maverick also explained its plans to invalidate their patent because the invention claimed in the competitor's patent was actually invented by Maverick several years earlier. Because Maverick made the invention well before the filing date of its competitor's patent, Maverick planned to file appropriate paperwork to initiate an interference proceeding in the U.S. patent office. The anticipated outcome of this proceeding would be both a declaration that Maverick was the first inventor of the invention and an assignment of all relevant patent claims to Maverick.

The all-out attack had the desired outcome. After receiving this communication, Maverick's accuser agreed to talk and the parties quickly came to terms on a settlement. Even with a quick settlement, Maverick incurred significant legal fees. But it could have been worse—a lot worse. If Maverick hadn't had the funds to raise its defenses, the company could have quickly gone out of business, even if it eventually won at trial. Imagine how many mountain bikes a company would need to sell to make a million dollar profit. Even worse, while embroiled in a lawsuit, no manufacturer would be willing to keep doing business with Maverick. There are just too many other designs for them to choose from.

Chapter Twenty-Two
HOW DO WE FIX THIS?

What few people realize is that the inventive spirit that reinvigorated America during the nineteenth century is again trying to take root—at a time when it is desperately needed. The parallels are striking. Although today we talk in terms of the Internet, superconductors, cellular telephones, and smart bombs, you could just as easily replace these technologies with those of the 1830s.

That seems impossible, but it's true. The decade we just experienced was little more than a repeat of what happened in the 1830s. The only difference is what came out of the 1830s has yet to find is equal in the world's history.

America's frenzy over the blossoming of the Internet brought about a dot-com craze that seems unprecedented—until it is compared with America's "Rubber Fever." It too made and lost fortunes, while also laying the groundwork for a massive technological boom.

One benefit that survived the dot-com disaster was the immense telecommunications infrastructure that it put in place. Millions of miles of high-speed fiber optic cables were strung over the entire globe. The same thing happened in the 1830s as Morse's experiment with the telegraph finally took hold and Congress appropriated $30,000 for a 40-mile line from Washington, D.C. to Baltimore. In just a few years, telegraph lines found their way to nearly every major city in the world. People could now communicate in near real time—no matter the distance.

The race to lay cable wasn't the only race. America was in an arms race as well—stemming from America's clash with Spain over Florida, then with Mexico over the Mexican territories. As it does now, the U.S. army sought for a technological advantage. And they found it in Colt and his revolver. The nuclear arms race between the U.S. and the former Soviet Union is just one example of how similar we are to former generations.

There was also a food revolution. Then it was wheat driving the need for the mechanical reaper. Today it involves corn and the efforts to make high fructose corn syrup or ethanol.

In so many ways, the 1830s was similar to the modern era, save one: the drive to invent. The great inventors of the 1830s spent decades in total dedication to invent their ideas, living on next to nothing and barely eking out a living. Today's inventors have nine-to-five jobs in research labs, with perks like gourmet cafeterias, game rooms, and high-tech gymnasiums. For most, all their forebears had was just guts and determination, and a desire to forge ahead.

The incentive for these early innovators wasn't just the satisfaction of inventing. Mostly, it was the financial reward, a way for them to secure their futures. And it was the patent system, more than anything else, that gave them the faith that if they invented, they would reap the financial rewards.

The obvious impact of the patent system on the U.S. economy was understood the world over. Seeing America's propulsion onto the international stage, the Japanese government sent a contingency to the U.S. to study the patent system. When asked why the Japanese people wanted a patent office, the special commissioner replied, "What is it that makes the United States such a great nation? And we investigated and found that it was patents, and we will have patents."

Today we still have a patent office, but where are the inventions? Where is our technology revolution? Where are our alternative fuels? When do we say goodbye to the Middle East? Where are the McCormicks and the Goodyears of the world?

We need to invite them back, just as we did when we first extended the invitation in 1836, a time of unprecedented innovation in America.

The problem with today's culture is that Americans are unwilling to dedicate twenty years to inventing a technology when they have no guarantee that their ideas will be protected. Not so a century and a half ago.

Morse is an extreme example of what it took to succeed. Upon returning from Europe, he told his brothers about his extraordinary idea to communicate over a wire. While still clinging to the hope of being a great painter, he continued to tinker with the idea for another five years. When his painting career ended, he struggled for years to perfect his telegraph. His one token of triumph was the granting of his patent. Yet even with patent in hand, it took the help of Samuel Colt and the head of the patent office to secure him a $30,000 grant from Congress to lay his first line. When the infringers copied his telegraph, it took a team of lawyers to stop them. Yet the patent system *did* work. The telegraph was both the most financially successful and society changing invention of the nineteenth century, and it eventually paid Morse handsomely. The same could be said for Colt's revolver and Goodyear's vulcanized rubber.

If it was that difficult to introduce a groundbreaking technology in the mid nineteenth century when the patent office heavily favored inventors, consider what it's like today when the patent office and the courts are so uncooperative. An inventor like Morse has virtually no chance.

So what can be done to bring back America's golden days of invention —a time when we invented twice what we do today?

A good start would be to re-create the inventive environment of a bygone era. To bring back America's innovative spirit, we must create a more efficient, equitable process. Inventors can't keep waiting five years and shelling out $30,000 for a patent—if they can even get one—then expect to pay $5 million to enforce their patent or defend against someone else's patent. That means figuring out what the modern equivalent to the nineteenth-century approach might be.

It's not impossible. Just look to what is happening with smart phone applications or "apps" and apply that to every technology field. Although these apps are wildly successful, it's only because the amount of investment needed to create one is so low the entire industry has developed outside the patent system. If our patent system efficiently protected inventions that do take significant capital expenditures and still gave room for competition, you'd see everyone inventing. And they'd be inventing the technologies we desperately need, such as more efficient forms of transportation and clean energy.

What we need is a role reversal, where inventors are those outside the walls of corporate America, like those who came up with the telegraph and the airplane. Under our current patent system, only the rich can play.

We've now had 220 years of experience with patents. Thomas Jefferson's idea was to use history as a guide for future patent laws. This was done in 1836 and was immensely successful. But now we've veered off course, and with more than a century of experience since 1836, perhaps it's time for us to step back, evaluate what has happened to our patent system and rewrite the patent statute. In large part, we can simply return to 1836 when Senator Ruggles of Maine took a fresh look at forty years worth of mistakes—promulgated by Jefferson's refusal to examine one more patent application—and created a vastly improved patent act. But we can even do better that Senator Ruggles. We can simplify even more, eliminating the undesirable aspects of the 1836 Act.

It's clear that the biggest reason for our inventive demise is America's complex, over-legislated, under-administered patent system. Inventors need funding to bring their ideas to market, then need an efficient way to keep copiers from undermining these investments. To make that happen they need to protect their ideas—and for that, they need a patent. Colt paid around $30 to obtain a patent that was granted in only a few months, then raised around $230,000 in working capital. Today, that is not even within the realm of possibility. Goodyear was ridiculed for paying Daniel Webster $15,000 to enforce his rubber patent, yet today he could expect to pay several hundred times that amount.

A good part of today's complexities have resulted from judicial decisions trying to curb the power of patents, patents like those issued to the Wright Brothers. Some might think that a strong patent system will hinder innovation, especially if new improvements can't be introduced into the market because of a blocking patent. The reaper and sewing machine industries show that simply isn't true.

Although the reaper and the sewing machine industries developed in parallel, both generating hundreds of patents, and the ensuing litigation proceeded on vastly different fronts, both came to an efficient resolution. With the reaper, infringement cases were relatively limited in number, most likely because the reaper gained commercial acceptance after McCormick's original patent had expired. Most of the reaper litigation involved enhancements to the original reaper design. Perhaps even more important, McCormick's request for a patent extension on his improved design was denied, permitting the entire reaper market to incorporate McCormick's features into their own reapers.

Another way the early patent laws succeeded was by encouraging a diversity of new ideas within a given industry. With the reaper industry, farmers made so many improvements that manufacturers had a wide assortment of new technologies to choose from, thereby eliminating the need to infringe on a competitor's design. This allowed humble farmers to invent—not massive companies. Although initially crude, the reaper design rapidly refined itself, with hundreds of farmers and mechanics contributing to its ever-evolving design. The potential for being awarded a valuable patent whipped innovators into a frenzy. Although this produced its fair share of overlapping patent rights and numerous patent infringement suits, the cost of these cases was on the order of a few thousand dollars. It allowed the new market entrants to successfully fend off patent infringement charges and allowed them to compete.

With the sewing machine, on the other hand, nobody could make a viable machine without infringing on somebody's patent because there were so many core patents. The existence of such an arsenal of patents resulted in industry-wide litigation that crippled nearly every company. On the verge of annihilation, the sewing machine companies ended up pooling their patents and coming to what was in essence a gentlemen's agreement, thereby rewarding patent holders according to the perceived market value of their patents. Although Howe did receive an extension of his sewing machine patent on the lock stitch, the patent system was well suited to permit the parties work out a compromise. With today's complexities, that would be unlikely.

It is clear that to bring back America's innovative spirit, we need a more efficient, equitable process. We need a modern equivalent to the nineteenth-century approach. Here is what is needed:

Return the patent models

Until the 1880s, inventors were required to provide patent models to prove that they had actually created their ideas. In part, this requirement was abandoned because drawings were logistically so much easier to file and store. Today we too can easily produce patent drawings, but that sidesteps the real issue. Models aren't necessarily needed in order to *explain* the invention, but rather to prove that inventors really have *invented* something; otherwise inventors can patent the merest whim of an idea.

Today, this happens all the time. Clever patent attorneys lodge patent applications on ideas they speculate will be developed in future years, then sit on their applications until the opportunity arises to attack an unsuspecting infringer. The patents they assert are typically "paper patents," ideas scribbled on paper but never actually built.

The concept of these kinds of "paper patents" is nothing new. The most famous early case was that of the Selden automobile patent asserted against Henry Ford. Ford spent much of his time and treasure fending off Selden and the Association of Licensed Automobile Manufacturers. The problem stemmed from the fact that Selden never built a working automobile before filing his application, but was a clever patent attorney who manipulated the patent system for sixteen years until the automobile business became a commercial reality. Although he eventually did submit a patent model, it was a non-functioning model, a mere toy car. All this costly litigation could have been avoided if Selden had been required to build his car before filing a patent application—a patent office rule that was eliminated the year before Selden filed his application.

The same could be said for Bell's telephone patent. He too did not build a working telephone before filing his application. Evidence suggests to some historians that Bell lifted language from Elisha Gray's patent application that was filed the same day. If Bell had had to file a model upon filing his application, his patent would never have been granted.

Many of today's critics cry for patent reform because they feel that so many illegitimate patents have issued that they block important technologies from coming to market. Bringing back the models squarely addresses this issue by eliminating paper patents and most of the dubious litigation around so-called "patent trolls" who obtain patents for the sole purpose of obtaining patent royalties from others. These "trolls" prefer to be called "non-practicing entities," because they have no underlying business. In most cases, there is also no underlying technology.

The use of paper patents, regardless of who enforces them, has clogged our courts with dubious litigation, none of which serves to advance technology. Instead, it merely lines the pockets of those who know how to manipulate the patent system. The number of current patent trolls enforcing patents costs companies billions of dollars.

The granting of a patent right is perhaps the strongest right afforded by our government. We can't be giving this significant right away based on a poorly crafted document of two pages. The amount of contribution should be commensurate with the magnitude of the granted right. We've got to stop granting rights that end up costing defendants millions of dollars without a quid pro quo disclosure. Although individual inventors may be disadvantaged by the requirement for a model (because of the increased expense and difficulty in reducing their ideas to practice), the harm to society is much greater without this evidence. Why give someone a twenty-year monopoly for just jotting down the shadow of an idea and then turn him loose in our litigation system?

Critics may argue that the number of patent applications would drastically fall if a policy requiring a reduction to practice were implemented. They may say that this will curtail innovation. After all, it will disadvantage under-funded inventors who are now required to build their idea before filing for patent protection.

Perhaps, but it will also have the effect of stabilizing the patent system, thus increasing the amount of innovation. After certainty returns, more people will begin inventing. Even more important, the patent laws could be greatly simplified to allow patents to issue more easily. Under current laws, obtaining patents is so difficult (primarily because of the law on obviousness discussed below) that more people would be encouraged to invent and patent. What this means is that the cost of building the invention will be greatly offset by the cost savings obtained by quickly moving the patent application through the patent office. And, historically, when most inventors were from the working class, the requirement to submit a patent model didn't stifle innovation. Today, the requirement wouldn't be to make a model, but would be to make it the real thing. Unlike former times, inventors wouldn't need to locate a model shop, but just build their invention. Ultimately, this requirement is a way for the inventor to show some level of commitment—not by haggling with the examiner for years as is now the case, but by making the inventor show how his idea works.

If the number of patent applications will fall, doesn't this contradict the goal of increasing innovation? Frankly, it would be a good thing if fewer patent applications were filed in the short run. It would give the patent office some breathing room, a chance to get caught up. The past two patent commissioners have repeatedly stated that one of the most significant issues facing the patent office is the backlog of unexamined applications. When a patent application is filed, it sits in a queue until reaching the top of the stack. Currently, it can take three to five years for a patent to be examined. If people are forced to file better crafted patent applications, they will file fewer and this backlog would quickly dwindle. It would hasten the application's journey through the patent office because the examiner could actually see the idea embodied, rather than spend time wading though vague language in the patent specification. And it would reduce the number of "enablement" rejections—a patent office requirement that the applicant described in words why an invention will work. This would also be true when enforcing a patent because the parties would not need to argue over whether the patent sufficiently described or "enabled" the invention. With a smaller backlog, examiners could do a better job searching for prior art and examining applications. And, with this new approach, the number of patents with questionable validity would drop drastically—a much better approach to improving quality than rejecting most applications outright as is now in vogue.

When the patent system is fixed, when the patent office can grant a patent in less than a year, and when those patents are quality patents, people will begin to invent and more applications will inevitably follow. With better patents coming out of the patent office, the number of patents litigated will also decrease. Patent rights, now clearly delineated, will stand a better chance of being respected. If a competitor can define with certainty what a competitor's patent covers, it can design other products that do not infringe, or else negotiate a license. In the event that a patent dispute is litigated, models will provide visual evidence to the jury and help the judge to define the claims, just as the Brayton model did during the Ford litigation. The same thing happened in the McCormick case, where the judge used McCormick's model to quickly determine that Manny's reaper didn't have the claimed raker.

What if inventors were once again required to submit proof of their reduction to practice? How would it work? Obviously, submitting models is problematic. In 1880, the patent office couldn't deal with the thousands of models it received annually. Today, that number would be in the hundreds of thousands.

One simple solution is to document proof of construction on video. Today, patent applications can be filed by mail or electronically. For those filing by post, a DVD could be submitted along with the application. If filed electronically, a movie file could be uploaded to the patent office web site. After eighteen months, not only would the patent application be published (as is currently the case), but the video could be made available for public inspection. If the patent office wanted to get creative, they could even post them on YouTube.

Of course, there will be cases where a video won't suffice, as with the workings of many nanotech inventions, such as a fluidic valve, which is a small circular opening of around 100 microns that stops fluid flow by surface tension forces. For these, the valve could be shown working in an overall system, and a blown-up model of the idea could be submitted. For chemical compounds or processes, videos of practical applications could be submitted. Although some accommodations may need to be made for certain circumstances, with today's technology, submitting proof of actual reduction to practice is a relatively simple endeavor, and one that is essential to preserving the integrity of the patent system.

Many modern inventions are for software. It would be relatively easy to have the inventors produce their code. That is the simplest way to prove they have created their invention.

Of course, the patent office is opposed to this. When asked about bringing back a modern-day equivalent to models, the current Commissioner shouted, "No! Where would I put them?"

He obviously missed the point. We don't need any three-dimensional models. A simple video clip would suffice. And for that, all the patent office needs is a server.

Abolish the *obviousness* standard and the *doctrine of equivalents*

Abolishing both of these requirements will stop the fighting over how *meritorious* an invention must be before being awarded a patent, while concretely defining the bounds of a patent claim.

In the nineteenth century, patents were awarded to inventions that were "new." The problem arose in determining what was new. The original test was quite simple: Compare the idea described in the patent with existing ideas, as described in a piece of literature or as shown in a previous patent model, and if the idea was different in any respect, the invention was new.

Over time, a theory began to develop that since patents were so important and powerful, the idea had to be more than just "new." An inventor had to show something "extra-special" about his invention before being awarded a multi-year monopoly. The real reason behind this was that the courts thought too many patents were getting through the patent office and wanted inventors to show that they had made significant, non-trivial advancements over the current state of technology. If the patent applicant couldn't do this, the claim would be rejected for being "obvious." In other words, an idea was obvious to come up with unless it was "truly inventive." Fine, until you try to determine what is "inventive" as opposed to just "new."

In reality, no test will work. The current situation in the patent office is so dire that most of this country's greatest inventors, like Edison and Morse to name a few, would have had their patent applications denied. For Morse, the case would not have even been close. Even Justice Taney admitted that Morse had cobbled together old ideas.

The immediate benefit of eliminating this requirement would be to drastically reduce current backlogs in the patent office. Whereas it now takes three to five years to get a patent, only a few decades ago it was a matter of months.

Some will argue that certain inventions should not be patentable for various social reasons and the obviousness standard serves as a gatekeeper to prevent patents from issuing on those technologies. That misses the point. If you want to protect a class of practitioners, like doctors, just have Congress pass a law saying you can't sue a doctor for patent infringement and collect damages. But don't make the patent office the arbiter of what is obvious and what isn't. That just stifles innovation.

The quid pro quo for eliminating the obviousness requirement is to also drop the doctrine of equivalents. Put another way, the flip side of the "obviousness" test is the "doctrine of equivalents." Patent holders often cite the doctrine of equivalents to argue that an alleged infringer is violating a patent claim even if the infringer isn't doing exactly what is spelled out in the claim. The doctrine of equivalents allows a patent holder to expand the scope of the patent so that the infringer cannot make a trivial change to the product and argue that it falls outside the scope of the patent claims. In essence, this is the argument made by the Wright Brothers in their case against Glen Curtiss over the use of ailerons. Although the Wright Brothers won this argument, McCormick in his battle with Manny did not because the doctrine of equivalents didn't exist in 1850.

By way of illustration, consider the bicycle described in the introduction, with a sensor that sends wireless signals to a microprocessor to determine the power expended by the rider. Rather than use a wireless sensor in the bicycle seat, the infringer may decide to place a wire that runs alongside the brake cable. Using the doctrine of equivalents, the patent holder could argue that, while the infringer's invention doesn't exactly do what the claim requires—sending a wireless signal from the sensor to the microprocessor—it is so close that it should still be considered an infringement.

It's easy to see the problem with this doctrine. How does a competitor know how far he must distance his competing device from the claims of the patent in order to avoid infringement? That is one reason why patent infringement cases end up costing millions of dollars.

Eliminating the doctrine of equivalents dovetails with the abandonment of the non-obviousness requirement. How does the doctrine of equivalents go hand in hand with the obviousness standard? Perhaps some broader patents may issue if the obviousness standard were eliminated (because patents need to be only different than what has been known before), but if the doctrine of equivalents is abolished, the claim scope cannot be expanded during litigation. In other words, if broader patents are granted by eliminating the obviousness standard, the doctrine of equivalents should be abolished to keep patent claims narrow. You can't have one standard and not the other. But having both simply adds unnecessary complexity, cost, and uncertainty.

Cut the current twenty-year patent term in half

Today, a patent may be enforced for twenty years from the "priority date." This is the earliest date from which the application claims priority, and is usually just the filing date. In many cases, when the patent office takes too long to issue a patent, the twenty-year term can be further extended.

With how quickly today's technology advances, there is no need for a twenty-year monopoly. Under the first patent act, the term was only fourteen years. A better term would be ten years, as adopted by the ancient Venetians. If not that, at least go back to the original fourteen-year term laid down by America's founders.

The major reason why the fourteen-year term was extended was because some inventors were unable to commercialize their ideas in fourteen years and began to complain. It didn't take long for Congress to change the patent laws so that inventors could petition for a "patent extension." This enabled inventors to tack on seven years of patent life if the patent holder was unable to commercialize his idea after diligent efforts. In essence, these extensions prolonged the monopoly and reduced competition. Morse, Colt, Goodyear, and Howe all got them, and they all reaped enormous profits from them while keeping out any competition. Eventually, Congress decided simply to give everyone a twenty-year term.

It is clear that if patent terms were significantly reduced, companies would be unable to sit on their technologies and we would be able to get rid of the problems America experienced with patent term extensions. Few realize that many corporations hold patent strategy meetings where the main topic of discussion is to calculate the expiration dates of their main patents, then map out when they should start funneling additional money into R&D, typically within about 3-5 years of their patents expiring. Until then, they are content to just sit on their technology and be prepared to enforce their patents.

Colt and McCormick are good examples of what happens when patent terms are limited. For Colt, extra patent life was nullified when he bribed members of Congress when asking for his second extension. When his patent expired, numerous competitors entered, forcing Colt's company to continue innovating as well as streamlining his manufacturing process so that he could continue to dominate the market. If his patent had been extended, Colt could have continued his monopoly well past the Civil War.

McCormick, denied his patent term extension, also had to innovate to stay competitive, and he used the American farmer to do this, stimulating economic development throughout the expanding West.

There are other reasons for reducing patent terms. For example, if the obviousness requirement were eliminated, patents would issue much faster, perhaps within a year as they did during the nineteenth century. Similarly, if inventors were required to build their inventions prior to filing, the product would be ready to commercialize and there would be no need for extra patent life while the inventor struggled to build a commercial product. Finally, a principle known as Moore's law holds that technology doubles every two years. A twenty-year term discourages the kind of rapid innovation that would otherwise naturally occur. At the same time, ten years gives investors plenty of time to earn a return on their investment.

Some, typically the drug companies, argue that they need to have extra life to compensate for the millions or even billions of dollars that it takes to discover and launch a successful drug. Perhaps they should have their own deal, but for the rest of the inventing world, ten years is plenty.

Curtail the continuation practice

A major glitch in today's patent laws is the practice of allowing "continuation applications," where applicants can get one patent, then go back in hopes of getting a broader claim in an additional patent. The only limit to how many times an applicant can take another bite of the same apple is the patent term which limits all related patents to a term of twenty years. The twenty year date is calculated from the date when the first application of the group was filed. Big corporations with large bankrolls live by this practice. The problem is that for twenty years competitors have no idea what claims will pop up from the patent office because any number of continuation applications can be filed during the twenty year term. The ability to file an unlimited number of continuations is arguably the single biggest cause of the backlog in the U.S. Patent Office. No other patent office in the world allows continuation applications.

If applicants don't get the claims they want the first time, they want the ability to keep asking the patent office for more claims until they do—a so-called "fair treatment" argument. This argument does have some merit, especially if the applicant draws an unreasonable examiner who latches onto the flimsiest of "obviousness" rejections. But if the obviousness standard were eliminated so that broader patents could be issued the first time, there would be no need for continuation applications. In turn, this would increase the certainty that an idea will be appropriately protected. Also, if only one patent is granted for each invention, competitors can get to work coming up with alternative ways to build technologies or else negotiate licenses to the original patent, as opposed to doing nothing because producing a competing product is just too risky when you don't know what other patents may emerge.

The automobile litigation is a prime example. Any system with rules is subject to manipulation, and George Selden was a master at manipulating patent rules. He managed to keep his patent application secret for sixteen years until he obtained the claims he wanted, then asserted his patent against the dozens of newly formed automakers. Selden's ability to alter his claims over a period of several years had the ultimate effect of crippling the entire automotive industry.

Problems with continuations are similar to those that occurred with the early practices of reissuing patents and granting patent term extensions: namely, they created uncertainty as to when competitors could enter the market. Nowhere is this more evident than with Morse and Goodyear. Their competitors (primarily O'Reilly and Day) were kept at bay for years. Even using creative accounting, Morse and Goodyear should not have been able to extend their patents, yet because they were sentimental favorites, gaining the sympathy of the patent office because of their hardships, they each managed to get themselves another seven years of patent life. Meanwhile, those who assumed that Morse and Goodyear were not entitled to a patent term extension were now faced with how to compete in a patent-dominated market for another seven years.

For the sewing machine industry, competitors never knew how long they had to keep paying into the Combination. Howe's patent should have expired in 1860, but he managed to convince Congress to extend the patent to 1867. When he tried a second time for another seven-year term, the request was denied.

The reissue practice, still in existence today, creates similar uncertainties because a patent holder can ask the patent office to change his claims after his patent has issued. Colt's attorney, Dickerson, was able to increase the scope of Colt's patent when he reissued Colt's patent right before suing Massachusetts Arms. This too is a doctrine that can be eliminated along with the continuation practice.

Go to a first-to-file system

The telephone scandal created by Alexander Bell when he filed his patent application should be enough to convince anyone that patents should be granted to the first person who files his patent application with the patent office, not the one believed to have invented first. In other words, if two inventors file on the same idea, the one who files his patent application first is awarded the patent. First to invent is now determined in an "interference" proceeding in the patent office, and takes years and hundreds of thousands of dollars to determine. Eliminating interferences would free up additional patent office resources for examining patent applications. Congress recently passed legislation adopting a first-to-file system. However, it is not a true first-to-file system in that a person can publish his idea first, then wait a year to file his application.

Bell wasn't the only one who argued he was first to invent. Morse did so twice to keep out alternative telegraph designs. Horace Day tried it too, but Henry Ellsworth discovered the ruse and denied his request. And Singer tried it with Howe, but the judge denied his "rusty claims." These could have all been avoided with a first-to-file system.

Some may think these measures are extreme, but others have recognized the benefits of a streamlined patent system. The Australian patent office now has an "Innovation Patent" that is similar in many respects to the simplified patent system presented above. To receive an Innovation Patent, the innovation must be new. This means that the invention must not have been publicly disclosed or already known. This is the novelty requirement required by all patents, U.S. and worldwide.

However, the innovation is not tested against an obviousness requirement, referred to internationally as "inventive step." Instead, it must only be "innovative." The test for innovation is whether the claimed invention makes a "substantial contribution" over the prior art, a much lower standard than proving an idea is "non-obvious." These innovation patents issue within months, and after they are certified they can be enforced in court. Similar to the proposal outlined above, these innovation patents have a an eight-year term. It is still too early to tell what impact these innovation patents will have on invention in Australia, yet word is quickly spreading.

Nevertheless, history has proven that a simplified patent system will spur innovation in a remarkable way. It will foster a climate that is conducive for innovation, just as it did during the careers of Colt, Goodyear, Morse, and McCormick.

Is it realistic?

One thing is for certain: piecemeal legislation intended to address narrow issues raised by those with money and political clout will only make matters worse. The patent system needs a serious overhaul, a drastic simplification. Doing so would renew America's spirit of innovation and open a new chapter in American history. Whether Congress has the fortitude to deliver is another story.

ABOUT THE AUTHOR

Darin Gibby is a patent attorney with Kilpatrick Townsend and has nearly twenty years of experience in obtaining patents on hundreds of inventions from the latest mountain bikes to life-saving cardiac equipment. He has built IP portfolios for numerous Fortune 500 companies and has monetized patents on a range of products from computer disk drives to in-line skates. He is a sought-after speaker on IP issues at businesses, colleges and technology forums. He currently resides in Colorado with his wife, Robin, and four children.

RESOURCES

Visit

www.DarinGibby.com

to learn more about America's famous patent battles,
view photos of old patent models, and much more.

REFERENCES

Books

_____., *American Enterprise: Nineteenth-Century Patent Models*, Cooper-Hewitt Museum, New York, N.Y., 1984.

_____., *The Crystal Palace Exhibition*, Dover Publications, Inc., New York, N.Y 1970.

Dobyns, Kenneth W., *The Patent Office Pony*, Sergeant Kirkland's Museum and Historical Society, Inc., Fredericksburg, VA., 1997.

Evans, Harold, *They Made America*, Back Bay Books, New York, N.Y., 2004.

Ffrench, Yvonne, *The Great Exhibition:1851*, The Harvill Press, London 1950.

Goodwin, Doris Kearns, *Team of Rivals: The Political Genius of Abraham Lincoln*, Simon & Schuster, New York, N.Y., 2005.

Green, Constance McL., *Eli Whitney and the Birth of American Technology*, Longman, New York, N.Y. 1956.

Greenleaf, William, *Monopoly on Wheels*, Wayne State University Press, Detroit, Mich., 1961.

Houze, Herbert G., *Samuel Colt: Arms, Art, and Invention*, Yale University Press, New Haven, Conn., 2006.

239

Howard, Fred, *Wilbur and Orville: a biography of the Wright Brothers*, Dover Publications, Inc., Mineola, N.Y.,1998.

Kahn, B. Zorina, *The Democratization of Invention: Patents and Copyrights in American Economic Development, 1790-1920*, Cambridge University Press, New York, N.Y. 2005.

Shulman, Seth, *The Telephone Gambit*, W.W. Morton & Company, 2008.

Slack, Charles, *Noble Obsession*, TEXERE LLC, New York, N.Y., 2002.

Silverman, Kenneth, *Lightning Man*, Da Capo Press, Cambridge, Mass., 2004.

Standage, Tom, *The Victorian Internet*, Walker and Company, New York, N.Y., 1998.

Publications, Newspapers, Cases and Other Articles

Lundeberg, Philip K., *Samuel Colt's Submarine Battery*, Smithsonian Institution Press, Washington D.C., 1974.

Mossoff, Adam, *A stitch in Time: The Rise and Fall of the Sewing Machine Patent Thicket*, Arizona Law Review, Vol. 53, pp. 165-211, 2011.

MacGregor, Donald, "The Passing of Uncle Sam's Old Curiosity Shop," *The Mentor*, September, 1925.

Seabrook, John, "The Flash of Genius," *The New Yorker*, January 11, 1993.

"A Snow Storm," *The New York Times*, November 1, 1851.

"Colt's Revolvers," *The New York Times*, October 14, 1852.

"Colt's Revolvers," *The New York Times*, October 19, 1852.

"The Great India Rubber Case," *The New York Times*, October 22, 1851.

"The Great India-Rubber Case—New-Jersey Legislature," *The New York Times*, March 24, 1852.

"The Great India Rubber Case," *The New York Times*, March 30, 1852.

"Webster's Argument—Goodyear vs. Day," *The New York Times*, May 14, 1852.

Colt v. Massachusetts Arms Co., 6 F.Cas. 161 (D. Mass. 1851).

Cyrus H. McCormick v. Waite Talcott et al., 61 U.S. 402 (1858).

Henry O'Reilly et al. v. Samuel F.B. Morse et al., 56 U.S. 62 (1853).

Goodyear v. Day, 10 F.Cas. 678 (D. N.J. 1852).

McCormick v. Manny et al., 15 F. Cas. 1314 (N.D. Ill. 1856)

Whitney v. Carter, Fess. Pat 130, 29 F.Cas. 1070 (D. GA 1810)

Wright Co. v. Herring-Curtiss Co. et al., 204 F. 597 (W.D. New York 1913).

BUY A SHARE OF THE FUTURE IN YOUR COMMUNITY

These certificates make great holiday, graduation and birthday gifts that can be personalized with the recipient's name. The cost of one S.H.A.R.E. or one square foot is $54.17. The personalized certificate is suitable for framing and will state the number of shares purchased and the amount of each share, as well as the recipient's name. The home that you participate in "building" will last for many years and will continue to grow in value.

Here is a sample SHARE certificate:

THIS CERTIFIES THAT
YOUR NAME HERE
HAS INVESTED IN A HOME FOR A DESERVING FAMILY
1985-2010
TWENTY-FIVE YEARS OF BUILDING FUTURES
IN OUR COMMUNITY ONE HOME AT A TIME
1200 SQUARE FOOT HOUSE @ $65,000 = $54.17 PER SQUARE FOOT
This certificate represents a tax deductible donation. It has no cash value.

YES, I WOULD LIKE TO HELP!

I support the work that Habitat for Humanity does and I want to be part of the excitement! As a donor, I will receive periodic updates on your construction activities but, more importantly, I know my gift will help a family in our community realize the dream of homeownership. I would like to SHARE in your efforts against substandard housing in my community! (Please print below)

PLEASE SEND ME _____ SHARES at $54.17 EACH = $ $_____

In Honor Of: _____

Occasion: (Circle One)   HOLIDAY   BIRTHDAY   ANNIVERSARY

    OTHER: _____

Address of Recipient: _____

Gift From: _____ *Donor Address:* _____

Donor Email: _____

I AM ENCLOSING A CHECK FOR $ $_____ PAYABLE TO HABITAT FOR HUMANITY OR PLEASE CHARGE MY VISA OR MASTERCARD *(CIRCLE ONE)*

Card Number _____ Expiration Date: _____

Name as it appears on Credit Card _____ Charge Amount $ _____

Signature _____

Billing Address _____

Telephone # Day _____ Eve _____

PLEASE NOTE: Your contribution is tax-deductible to the fullest extent allowed by law.
Habitat for Humanity • P.O. Box 1443 • Newport News, VA 23601 • 757-596-5553
www.HelpHabitatforHumanity.org

CPSIA information can be obtained at www.ICGtesting.com
Printed in the USA
LVOW071734151111

255122LV00001B/87/P